Decision Analysis through Modeling and Game Theory

This unique book presents decision analysis in the context of mathematical modeling and game theory. The author emphasizes and focuses on the model formulation and modeling-building skills required for decision analysis, as well as the technology to support the analysis.

The primary objective of *Decision Analysis through Modeling and Game Theory* is illustrative in nature. It sets the tone through the introduction to mathematical modeling. The text provides a process for formally thinking about the problem and illustrates many scenarios and illustrative examples.

These techniques and this approach center on the fact that (a) decision makers at all levels must be exposed to the tools and techniques available to help them in the decision process, (b) decision makers as well as analysts need to have and use technology to assist in the entire analysis process, (c) the interpretation and explanation of the results are crucial to understanding the strengths and limitations of modeling, and (d) the interpretation and use of sensitivity analysis is essential.

The book begins with a look at decision-making methods, including probability and statistics methods under risk of uncertainty. It moves to linear programming and multi-attribute decision-making methods with a discussion of proper weighting methods. Game theory is introduced through conflict games and zero-sum or constant-sum games. Nash equilibriums are next, followed by utility theory. Evolutionary stable strategies lead to Nash arbitration and cooperation methods and N-person methods presented for both total and partial conflict games.

Several real-life examples and case studies using game theory are used throughout. This book would be best suited for a senior-level course in mathematics, operations research, or graduate-level courses or decision modeling courses offered in business schools. The book will be of interest to departments offering mathematical modeling courses with any emphasis on modeling for decision-making.

Advances in Applied Mathematics

Series Editor: Daniel Zwillinger

Decision Analysis through Modeling and Game Theory

William P. Fox

CRC Press
Taylor & Francis Group
Boca Raton London New York

CRC Press is an imprint of the
Taylor & Francis Group, an **informa** business
A CHAPMAN & HALL BOOK

First edition published 2025
by CRC Press
2385 Executive Center Drive, Suite 320, Boca Raton, FL 33431

and by CRC Press
4 Park Square, Milton Park, Abingdon, Oxon, OX14 4RN

CRC Press is an imprint of Taylor & Francis Group, LLC

Library of Congress Cataloging-in-Publication Data
Names: Fox, William P., 1949- author.
Title: Decision Analysis through Modeling and Game Theory / William P. Fox.
Description: First edition. | Boca Raton, FL : CRC Press, 2025. | Series:
Advances in applied mathematics | Includes bibliographical references and index.
Identifiers: LCCN 2024021496 | ISBN 9781032721606 (hbk) | ISBN
9781032726915 (pbk) | ISBN 9781032726885 (ebk)
Subjects: LCSH: Decision making--Mathematical models. | Game theory.
Classification: LCC QA279.4 .F68 2025 | DDC 519.3--dc23/eng/20240628
LC record available at https://lccn.loc.gov/2024021496

ISBN: 9781032721606 (hbk)
ISBN: 9781032726915 (pbk)
ISBN: 9781032726885 (ebk)

DOI: 10.1201/9781032726885

Typeset in Palatino
by KnowledgeWorks Global Ltd.

This book is dedicated in memory of John Forbes Nash, Jr., whose influence during his visit to my class at the Naval Postgraduate School in 2009 had far reaching effects on my knowledge of game theory. Additionally, I want to thank Professor Frank R. Giordano for giving me the opportunity to teach game theory at the Naval Postgraduate School. Dedicated to my wife, Hamilton Dix-Fox.

Contents

Preface

Addressing the Current Needs

In recent years of teaching mathematical modeling for decision-making coupled with conducting applied mathematical modeling research, we have found that (a) decision makers at all levels must be exposed to the tools and techniques available to help them in the decision process, (b) decision makers as well as analysts need to have and use technology to assist in the entire analysis process, (c) the interpretation and explanation of the results are crucial to understanding the strengths and limitations of modeling, and (d) the interpretation and use of sensitivity analysis is essential. With this in mind, this book emphasizes and focuses on the model formulation and modeling-building skills required for decision analysis, as well as the technology to support the analysis.

Audience

This book would be best suited for a senior-level course in mathematics, operations research, or industrial engineering or graduate-level courses or decision modeling courses offered in business schools offering business analytics. The book *would be* of interest to mathematics departments that offer mathematical modeling courses with any emphasis on modeling for decision-making.

The following groups would benefit from using this book:

- Undergraduate students in quantitative methods courses in business, operations research, industrial engineering, management sciences, industrial engineering, or applied mathematics.
- Graduate students in discrete mathematical modeling courses covering topics from business, operations research, industrial engineering, management sciences, industrial engineering, or applied mathematics.
- Junior analysts who want a comprehensive review of decision-making topics.
- Practitioners desiring a reference book.

Objective

The primary objective of this book is illustrative in nature. It sets the tone in Chapter 1 through the introduction to mathematical modeling. In this chapter, we provide a process for formally thinking about the problem and illustrate many scenarios and illustrative examples. In these examples, we begin the setup of the solution process and also provide solutions in this chapter that we will present more in-depth in later chapters.

Based on many years of applied research and modeling, we have considered which techniques should be included or excluded in a book of this nature. Finally, we decided on the main techniques that we covered in our three-course sequence in mathematical modeling for decision-making in the Department of Defense Analysis and the Naval Postgraduate School. We feel these subjects have served and prepared our students well, as they have all gone on to become leaders and decision makers for our nation.

Organization

This book contains information that could easily be covered in a one- or two-semester course or a one-semester overview of topics such as a seminar class. This allows instructors the flexibility to pick and choose topics consistent with their courses and consistent with their current needs.

Decision theory covers decisions under risk, uncertainty, certainty multi-attributes, and conflict.

In Chapters 2–4, we present decision-making methods. We begin in Chapter 2 with a discussion of expected value in probability and statistics material and methods under risk and uncertainty. In Chapter 3, we discuss decisions under certainty with linear programming. In Chapter 4, we present multi-attribute decision-making methods as a discussion of weighting methods

Chapter 5 begins our discussion of game theory. We begin with total conflict games. These are also known as zero-sum or constant-sum games.

Chapter 6 covers partial conflict games where the values in the pay-off matrix do not sum either to 0 or the same constant. We begin with the classical games of the prisoner's dilemma and the game of chicken. We discuss pure strategy Nash equilibriums.

Chapter 7 discusses utility theory, which in the absence of real data, might be used to obtain the values in the pay-off matrix.

Chapter 8 discusses the Nash equilibrium, prudential, and counter-prudential strategies. The Pareto principle and Pareto optimal are illustrated.

Various mathematical methods are illustrated to find the Nash equilibrium when there are no pure strategy equilibriums.

Chapter 9 discusses evolutionary stable strategies with hawks and doves.

Chapter 10 is about strategic moves for partial conflict games. This is where communication between players in a game is allowed. We present first moves, threats, promises, and combinations of threats and promises.

Chapter 11 Nash arbitration and cooperation methods are discussed.

Chapter 12 N-person (actually three-person games) methods are presented for both total and partial conflict games.

Chapter 13 discusses games in extended form (game trees) with backward induction to solve. Several examples are provided.

Several real-life examples and case studies using game theory are used throughout the book.

In my career, I have published over 34 articles on game and decision theory applications as well as chapters in many textbooks. My students have enjoyed the many applications presented over the years. I believe one of the strengths of this text is the modeling and applications with game theory.

In this book, we cannot address every nuance in modeling real-world decisions. What we can do is provide a sample of models and possible appropriate techniques to obtain useful results. We can establish a process to "do modeling" and we can illustrate many examples of modeling and illustrate a technique in order to solve the problem. In the techniques chapters, we assume no or little background in mathematical modeling and spend a little time establishing the procedure before we return to providing examples and solution techniques.

The data used in the examples presented in this book are drawn from real, unclassified sources and are similar in nature and design to the actual data used in the real-world examples.

This book can apply to operations research analysts to allow them to see the range and types of problems that fit into specific mathematical techniques while understanding that all possible mathematics techniques that could be used were addressed. Because of space, we do leave out some important techniques such as differential equations.

This book also applies to decision makers. It shows the decision maker the wide range of applications of quantitative approaches to aid in the decision-making process. As we say in our modeling classes every day, mathematics does not tell what to do but it does provide insights and allows critical thinking into the decision-making process. In our discussion, we consider the mathematical modeling process as a framework for decision makers. This framework has four key elements: the formulation process, the solution process, the interpretation of the mathematical answer in the context of the actual problem, and sensitivity analysis. At every step along the way in the process, the decision maker should question procedures and techniques and ask for further explanations as well as assumptions used in the process. One major question could be, "did you use an appropriate technique" to obtain a solution and why were other techniques not considered or used? Another question could be "did you over simplify the process" so much that the solution does not really apply in this situation or were the assumptions made fundamental to even being able to solve the problem?

We thank all the mathematical modeling students who we have had over this time as well as all the colleagues who have taught mathematical modeling with us during this adventure. We are especially appreciative of the mentorship of Frank R. Giordano over the past 30-plus years.

William P. Fox
Professor Emeritus, Naval Postgraduate School
Visiting Professor, College of William and Mary

About the Author

Dr. William P. Fox is currently a visiting professor of computational operations research at the College of William and Mary. He is an emeritus professor in the Department of Defense Analysis at the Naval Postgraduate School and teaches a three-course sequence in mathematical modeling for decision-making. He earned his PhD in industrial engineering from Clemson University. He has taught at the United States Military Academy for 12 years until retiring and at Francis Marion University where he was the chair of mathematics for 8 years. He has many publications and scholarly activities including more than 20 books and 150 journal articles including:

Probability and Statistics for Engineering and the Sciences with Modeling using R (with Rodney X. Sturdivant), 2023, CRC Press; *Mathematical Modeling in the Age of the Pandemic*, 2021, CRC Press; *Advanced Problem Solving Using Maple: Applied Mathematics, Operations Research, Business Analytics, and Decision Analysis* (with William Bauldry), 2020, CRC Press; *Mathematical Modeling with Excel* (with Brian Albright), 2020, CRC Press; *Nonlinear Optimization: Models and Applications*, 2020, CRC Press; *Advanced Problem Solving with Maple: A First Course* (with William Bauldry), 2019 CRC Press; and *Mathematical Modeling for Business Analytics*, 2018, CRC Press.

1

Introduction to Decision Models

1.1 Overview of Decision-Making

In this book, we model decisions and study how to compute optimal solutions to the models representing those decisions. Our first distinction is whether the outcome or payoff to the decision-maker depends only on their decision or whether the outcome also depends upon a decision that one or more additional players (or decision-makers) make. For example, someone rolling a pair of dice may decide to bet on a total of 7. Whether they win or lose certainly depends on chance, but not on the decision of another player. Rather, the total sum of the dice follows a pattern which we can study. Then we can predict, over the long haul, how frequently we expect the total to be 7.

Suppose two countries are locked in an arms race and wish to disarm. If one of the countries disarms, its outcome depends not only on its decision but also on the decision of the second country: did the other country also disarm, or did it remain armed? As Figure 1.1 depicts, if the outcome depends upon only one player, we refer to those decisions as decision theory. If the outcome depends upon the decision of more than one player, we refer to those decision models as game theory. We now introduce these subjects by presenting illustrative examples, which we will solve later in the text that follows.

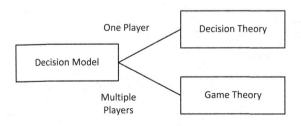

FIGURE 1.1
Illustration of decision theory (a) and game theory (b).

DOI: 10.1201/9781032726885-1

1.2 Decision Theory

The next distinction we make for decision theory is whether the process being modeled is deterministic or probabilistic. A process is deterministic if the same outcome results if the exact conditions are repeated. For example, deterministic models assume that known average rates with no random deviations are applied to large populations. For example, suppose there are 1,000 individuals and each has a 95% chance of surviving 1 year. Then we can be reasonably certain that 950 of them will indeed survive. While we give a precise definition later, for now, you can think of a probabilistic process as one where the outcome depends upon chance, even though the conditions are exactly repeated. For example, if you roll a pair of dice exactly the same, you do not expect to get the same result. Again, there is a pattern which we can study and use to predict the frequency with which each possible total occurs over the long haul (see Figure 1.2).

We further distinguish deterministic decision theory depending upon whether the decision model is unconstrained or constrained (see Figure 1.3).

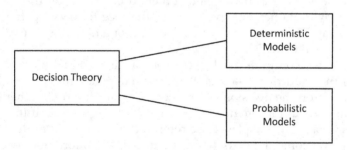

FIGURE 1.2
Deterministic decision theory.

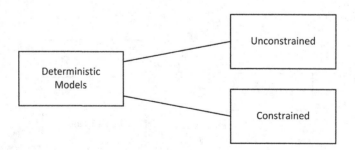

FIGURE 1.3
Constrained and unconstrained decisions.

We illustrate these concepts with examples.

Example 1.1: Unconstrained Deterministic Decision Theory

Suppose a decision analyst has modeled the total net profits P in thousands of dollars resulting from producing an amount x (in thousands) of an item as follows:

$$P = 60x - x^2 - 800 = (x - 20)(40 - x)$$

That is, if nothing is produced, the firm loses $800. If production is too large, then the $-x^2$ term indicates that the firm loses money at an increasing rate. Is there a range over which production is profitable? If so, what is the best or optimal production level? As the graph in Figure 1.4 indicates, the firm should produce 30,000 units to make $100,000. Further, the company remains profitable between 20,000 and 40,000 units.

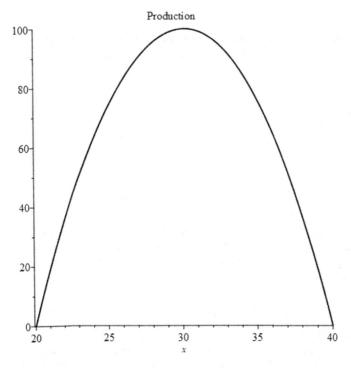

FIGURE 1.4
Graph of total net profit versus number of units produced.

Example 1.2: Constrained Deterministic Decision Theory

More typically, decisions are constrained. The classical problem in economics is to maximize profits subject to limitations on land, labor, and capital.

For example, airlines and oil companies would like to visit their designated cities and ports in such a manner to minimize the total distance traveled (need a better example than the traveling salesman problem). Let's consider an example.

A Montana farmer owns 55 acres of land available for planting. He is planning to plant wheat or corn. Each acre of wheat yields $200 in net profit, and each acre of corn yields $300 in net profit. The labor and fertilizer requirements per acre for wheat and corn are summarized in the following table. The farmer has 100 workers and 120 tons of fertilizer. Determine how many acres of wheat and corn need to be planted to maximize profit.

Summarizing the above data in a table, we have

	Wheat (per acre)	Corn (per acre)
Labor (Workers)	3	2
Fertilizer (tons)	2	4

Letting x represent the acres of wheat to plant, and y the acres of corn, we will learn to formulate the following decision model:

Maximize $Z = 200\,x + 300\,y$

Subject to:
$$3\,x + 2y \leq 100$$
$$2\,x + 4\,y \leq 120$$
$$x + y \leq 55$$
$$x \geq 0$$
$$y \geq 0$$

The above model is a linear program. We will learn how to model and solve decisions which can be represented by linear programs in Chapter 3 We will also use linear programming to find solutions to game theory models.

Example 1.3: Probabilistic Decision Theory

Consider a construction firm which is deciding whether to specialize in building high schools, elementary schools, or a combination of the two over the long haul. The construction firm must submit a bid which costs money and they may or may not be awarded the contract. If they bid on the high school, historical records indicate that they have a 20% chance of winning the contract and expect to make $50,000 net profit if they are awarded the contract. However, if they fail to get the contract, they will lose $1,000. The corresponding data for the elementary school is given in the following table.

High School	Elementary School
Win Contract: $50,000	Win Contract: $40,000 Profit
Lose Contract: $−1,000	Lose Contract: $−500
Probability of Award of Contract: 20%	Probability of Award of Contract:− 25%

Note that the high school is more lucrative because the profit increases to $50,000 over $40,000 if the contract is awarded.. However, the high school has a higher cost if the contract is not awarded and a lower probability of winning the contract. So, how do you make this decision? What are the possible objectives that the firm might have? We study various criteria that can be followed for making the decision under risk and uncertainty in Chapter 2.

Example 1.4: Probabilistic Decision Theory with Sequential Decisions

Many decision-making processes involve decisions which must be made sequentially. That is, the possible outcomes downstream depend upon one or more decisions that were made upstream. For example, you are in a Las Vegas Casino and have encountered the following game board, which is to be spun randomly (electronically), as illustrated in Figure 1.5.

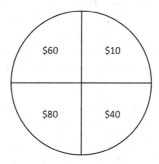

FIGURE 1.5
Electronic game.

Assuming that each outcome is equally likely, you compute the possible payoffs and probabilities as

Payoff ($)	Probability
10	1 in 4 (25%)
40	1 in 4 (25%)
60	1 in 4 (25%)
80	1 in 4 (25%)

You have a maximum of three spins but may stop after either the first or second spin. You have decided to play the game 100 times. Clearly, after the first spin, you would take $8 and stop, and you would spin again if you had a $0. But what about $4 and $6? And what is your criterion after the second spin? Clearly, after the third spin, if you get that far, you are stuck with whatever occurs. What you seek is an optimal decision-making strategy. That is,

> After Spin 1 Take:?
> After Spin 2 Take:?
> After Spin 3 Take: anything

If your goal is to maximize profit over the 100 games, what should you do? If you do want to make a profit, what is the maximum amount you should pay to play? We learn to model and solve sequential decisions in Section 3.5.

1.3 Game Theory: Total Conflict

Game theory is the study of decision where the outcome for the decision-maker depends not only on what he does but also on the decision of one or more additional players. We classify the games depending upon whether the conflict between the players is *total* or *partial*. We further classify games of total conflict depending upon whether the optimal strategies are pure or mixed, as illustrated in the following examples and in Figure 1.6.

Example 1.5: A Total Conflict Game with Pure Strategies

Suppose Large City is located near Small City. Now suppose a local neighborhood hardware chain such as Ace will locate a franchise in either Large City or Small City. Further, a "mega hardware store"

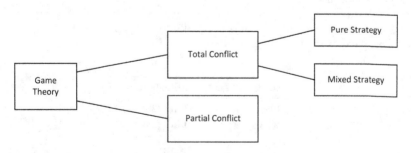

FIGURE 1.6
Game theory diagram.

franchise such as Home Depot is making the same decision – they will locate either in Large City or Small City. Each store has a "competitive advantage": Ace offers perhaps a convenient location, whereas Home Depot has a much larger inventory. Analysts have estimated the market shares as follows:

	Large City	Small City
Large City	60	68
Small City	52	60

That is, if both Ace and Home Depot are located in the same city, Large or Small, Home Depot gets 60% of the market. If Home Depot is located in Large City while Ace is located in Small City, Home Depot gets 68% of the market. But if Home Depot is located in Small City while Ace is in Large City, Home Depot gets only 52% of the market. Note that the profit that Home Depot makes depends not only on what they decide but also on Ace's decision. This is a *game* between Home Depot and Ace. Note also that the only way for Home Depot to get an additional 1% of the market is for Ace to lose 1%. That is, the game is *total conflict* since the sum of the market shares always total 100%. The corresponding figures for Ace are as follows.

	Large City	Small City
Large City	40	32
Small City	48	40

If you examine the situation for Home Depot, if Ace is located in Large City, Home Depot should be located in Large City (60 > 52). If Ace is located in Small City, Home Depot should be located in Large City (68 > 60). Regardless of what Ace does, Home Depot should be located in Large City. It is their *dominant strategy*. What strategy should Ace follow? As we see in the next example, a dominant or *pure strategy* is not always available, and the decision-makers must mix their strategies.

Example 1.6: Batter-Pitcher Duel: A Total Conflict Game with Mixed Strategies

In the game of baseball, a pitcher attempts to outwit the batter. Against a certain batter, his best strategy may be to throw all fastballs, while against a different batter; he might be better off throwing all curves. Finally, against a third hitter, his best strategy might be to mix fastballs and curves in some random fashion. But what is the optimal mix against each batter?

Consider the following table. A batter can either anticipate (guess) that the pitcher will throw a fastball or a curve. If he anticipates fast, he will either hit .300 or .200 depending upon whether the pitcher throws fast or curve.

If he guesses curve, he will either hit .100 or .500, again depending upon whether the pitcher throws fast or curve.

	Fast	Curve
Fast	.300	.200
Curve	.100	.500

Note in this case, if the batter guesses fast, the pitcher should throw curve. But if the batter guesses curve, the pitcher should throw fast. Unlike the previous example, the pitcher needs both strategies: fast and curve. But what is the optimal mix for the pitcher? Does the batter have the same optimal mix or a different mix? We learn to model and formulate total conflict models with mixed strategies in Chapter 5.

1.4 Game Theory: Partial Conflict

In the previous two examples, the conflict between the decision-makers was total in the sense that neither player could improve without hurting the other player. If that is not the case, we classify the game as partial conflict, as illustrated in the next example.

Example 1.7: Partial Conflict Game – the Prisoner's Dilemma

Consider two countries locked in an arms race: Country A and Country B. Let's begin by considering Country A, which can either remain armed or disarm. Its outcome depends upon whether Country B disarms or arms. Let's rank the outcomes as 4, 3, 2, and 1, with 4 being the best and 1 the worst.

	Disarm	Arm
Disarm	3	1
Arm	4	2

Arguably, Country A's best outcome (4) is if it is armed while Country B is disarmed. Its worst outcome (1) is if it is disarmed while Country B is armed. Now we must compare the situations where both countries are armed and both are disarmed. Arguably, both being disarmed is better than both being armed as there is less chance of an accident, and it is less expensive. So we give both being disarmed the 3 and both being armed the 2. The situation for Country B is similar, as illustrated in the following table.

	Disarm	Arm
Disarm	3	4
Arm	1	2

We now put the payoffs for both countries in a single table with Country B's payoffs listed second. That is, if Country A is disarmed and Country B is armed, the payoff (1, 4) indicates that Country A's payoff is a 1 while Country B's payoff is a 4.

	Disarm	Arm
Disarm	3,3	1,4
Arm	4,1	2,2

Now comes the dilemma. If both countries are armed, the payoffs are 2 for each country, their second worst outcome. If both disarm, both countries can improve to (3, 3), the second-best outcome for each country. Thus, the game is not total conflict. We will study what obstacles must be overcome in order to improve.

An important distinction in the study of partial conflict games is how the game is played: without communication, with communication, or with an arbiter, as indicated in Figure 1.7.

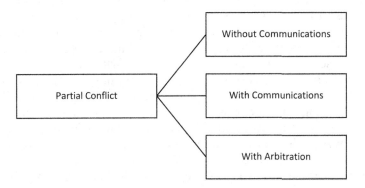

FIGURE 1.7
Partial conflict game diagram.

The communication may take the form of moving first and communicating to the other player that you have moved, making a threat to deter the opposing player from choosing a strategy that is adverse to you, or promising to choose a particular strategy if the opposing player chooses a strategy of your liking. Finally, we will study Nash Arbitration, a method for finding a negotiated fair solution based on the strategic strength of each player.

1.5 Mathematical Modeling of Decisions

Many large Fortune 500 companies have analysts to examine and build mathematical models to aid in decision-making. The decision modeling presented here crosses the lines of decision-making for business, industry, and government. Many examples are presented in this text from each of the sectors. Decision-making is often referred to as quantitative analysis, management science, and operation research. In this textbook, we will use a modeling approach to decision-making.

It is important to understand the process of modeling from problem identification to implementation. It is also important to know the strengths and limitations of these models. The correct use of good modeling tools and techniques results in solutions that are timely, useful, and easy to understand by those making the decisions.

Consider the importance of decision-making in such areas as business (B), industry (I), and government (G). BIG decision-making is essential to success at all levels. We do not encourage "shooting from the hip". We recommend good analysis for the decision-maker to examine and question in order to find the best alternative to choose or decision to make. So, why mathematical modeling?

A **mathematical model** is a description of a system using mathematical concepts and language. The process of developing a mathematical model is termed **mathematical modeling**. Mathematical models are used not only in the natural sciences (such as physics, biology, earth science, and meteorology) and engineering disciplines (such as computer science and artificial intelligence) but also in the social sciences (such as business, economics, psychology, sociology, and political science); physicists, engineers, statisticians, operations research analysts, and economists use mathematical models most extensively. A model may help to explain a system, to study the effects of different components, and to make *predictions* about behavior.

Mathematical models can take many forms, including but not limited to dynamical systems, statistical models, differential equations, or game theoretic models. These and other types of models can overlap, with a given model involving a variety of abstract structures. In general, mathematical models may include logical models, as far as logic is taken as a part of mathematics. In many cases, the quality of a scientific field depends on how well the mathematical models developed on the theoretical side agree with the results of repeatable experiments. Lack of agreement between theoretical mathematical models and experimental measurements often leads to important advances as better theories are developed.

1.5.1 Overview and the Process of Mathematical Modeling

1.5.1.1 The Modeling Process

In this chapter, we turn our attention to the process of modeling and examine many different scenarios in which mathematical modeling can play a role.

Mathematical modeling requires as much art as it does science. Thus, modeling is more of an *art* than a science. Modelers must be creative – willing to be more artistic or original in their approach to the problem. They must be inquisitive – questioning their assumptions, variables, and hypothesized relationships. Modelers must also think outside the box in order to analyze the models and their results. Modelers must ensure their models and results pass the common sense test. Science is very important, and understanding science enables one to be more creative in viewing and modeling a problem. Creativity is extremely advantageous in problem-solving with mathematical modeling.

To gain insight, we should consider one framework that will enable the modeler to address the largest number of problems. The key is that there is something *changing for which we want to know the effects*. We call this the **system** under analysis. The real-world system can be very complicated or very simplistic. This requires a process that allows for both types of real-world systems to be modeled within the same process.

Consider striking a golf ball with a golf club from a tee. Our first inclination is to use the equations about distance and velocity that we used in high school mathematics class. These equations are very simplistic and ignore many factors that could impact the fall of the ball, such as wind speed, air resistance, mass of the ball, and other factors. As we add more factors, we can improve the precision of the model. Adding these additional factors makes the model more realistic and more complicated to produce. Understanding this model might be a first start in building a model for such situations or similar situations, such as a bungee jumper or bridge swinger. These systems are similar for part of the model: the free fall portion has similar characteristics.

Figure 1.8 provides a closed-loop process for modeling. Given a real-world situation like the one above, we collect data in order to formulate a mathematical model. This mathematical model can be one we derive or select from a collection of already-built mathematical models, depending on the level of sophistication required. Then we analyze the model that we used and reach mathematical conclusions about it. Next, we interpret the model and either makes predict about what has occurred or offers explanation as to why something has occurred. Finally, we test our conclusion about the real-world

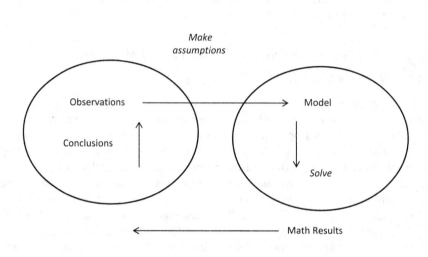

FIGURE 1.8
Modeling real-world systems with mathematics (see Albright et al., 2010).

system with new data. We may refine or improve the model to improve its ability to predict or explain the phenomena. We might even reformulate a new mathematical model.

1.5.1.1.1 *Mathematical Modeling*

We will build some mathematical models describing change in the real world. We will solve these models and analyze how good our resulting mathematical explanations and predictions are. The solution techniques that we employ in subsequent chapters take advantage of certain characteristics that the various models enjoy. Consequently, after building the models, we will **classify** the models based on their mathematical structure.

When we observe change, we are often interested in understanding why change occurs the way it does, perhaps to analyze the effects of different conditions, or perhaps to predict what will happen in the future. Often, a mathematical model can help us understand a behavior better while allowing us to experiment mathematically with different conditions. For our purposes, we will consider a mathematical model to be a mathematical construct designed to study a particular real-world system or behavior. The model allows us to use mathematical operations to reach mathematical conclusions about the model, as illustrated in Figure 1.8.

1.5.1.1.2 *Models and Real-World Systems*

A system is an assemblage of objects joined by some regular interaction or interdependence. Examples include sending a module to Mars, handling the United States debt, a fish population living in a lake, a TV-satellite orbiting the earth, delivering mail, and locations of service facilities – all are examples of a system. The person modeling is interested in understanding not only how a system works but also what interactions cause change and how sensitive the system is to changes in these inputs. Perhaps the person modeling is also interested in predicting or explaining what changes will occur in the system as well as when these changes might occur.

A possible basic technique used in constructing a mathematical model of some system is a combined mathematical-physical analysis. In this approach, we start with some known physical principles or reasonable assumptions about the system. Then we reason logically to obtain conclusions. Sometimes we have data and let help us come up with a reasonable model. Modelers must be open to many avenues to solve problems.

Figure 1.9 suggests how we can obtain real-world conclusions from a mathematical model. First, observations identify the factors that seem to be involved in the behavior of interest. Often we cannot consider, or even identify, all the relevant factors, so we make simplifying assumptions excluding some of them. Next, we conjecture tentative relationships among the identified factors we have retained, thereby creating a rough "model" of the behavior. We then apply mathematical reasoning that leads to conclusions about the model. These conclusions apply only to the model, and may or may not apply to the actual real-world system in question. Simplifications were made in constructing the model, and the observations upon which the model is based invariably contain errors and limitations. Thus, we must carefully account for these anomalies and test the conclusions of the model against

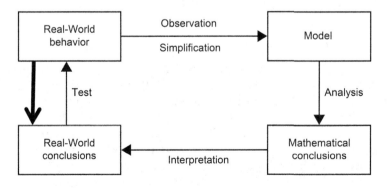

FIGURE 1.9
In reaching conclusions about a real-world behavior, the modeling process is a closed system (adapted from Giordano et al., 2013).

real-world observations. If the model is reasonably valid, we can then draw inferences about the real-world behavior from the conclusions drawn from the model. In summary, we have the following procedure for investigating real-world behavior:

Step 1. Observation the system or hypothesis the system (if one does yet exist), identify the key factors involved in the real-world behavior, simplify initially, and refine later as necessary.

Step 2. Conjecture or guess the possible relationships or inter-relationships among the factors and variables identified in Step 1.

Step 3. Solve the model.

Step 4. Interpret the mathematical conclusions in terms of the real-world system.

Step 5. Test the model conclusions against real-world observations – the common sense rule.

Step 6. Perform model testing or sensitivity analysis.

There are various kinds of models that we will introduce, as well as methods or techniques to solve these models, in the subsequent chapters. An efficient process would be to build a library of models and then be able to recognize various real-world situations to which they apply. Another task is to formulate and analyze new models. Still, another task is to learn to solve an equation or system in order to find more revealing or useful expressions relating to the variables. Through these activities, we hope to develop a strong sense of the mathematical aspects of the problem, its physical under-pinnings, and the powerful interplay between them.

Most models do simplify reality. Generally, models can only approxi-mate real-world behavior. Next, let's summarize a *process* for formulating a model.

1.5.1.2 Model Construction

Let's focus our attention on the process of model construction. An outline is presented as a procedure to help construct mathematical models. In the next section, we will illustrate this procedure with a few examples.

These nine steps are summarized in Figure 1.10, as modified from a six-step approach by Giordano et al. (2013). These steps act as a guide for think-ing about the problem and getting started in the modeling process.

Let's discuss each step in more depth.

Step 1. Understand the problem or the question asked.
Identifying the problem to study is usually difficult. In real life, no one walks up to you and hands you an equation to be solved.

Step 1. Understand the problem or the question asked.
Step 2. Make simplifying assumptions. Justify your assumptions.
Step 3. Define all variables and provide units.
Step 4. Construct a model.
Step 5. Solve and interpret the model.
Step 6. Verify the model.
Step 7. Identify the strengths and weaknesses of your model.
Step 8. Sensitivity Analysis or Model Testing of the model. Do the results pass the "common sense" test.
Step 9. Implement and maintain the model for future use.

FIGURE 1.10
Mathematical modeling process.

Usually, it is a comment like, "we need to make more money", or "we need to improve our efficiency". We need to be precise in our formulation of the mathematics to describe the situation.

Step 2. Make simplifying assumptions.

Start by brainstorming the situation. Make a list of as many factors, or variables, as you can. Realize we usually cannot capture all these factors influencing a problem. The task is simplified by reducing the number of factors under consideration. We do this by making simplifying assumptions about the factors, such as holding certain factors as constants. We might then examine to see if relationships exist between the remaining factors (or variables). Assuming simple relationships might reduce the complexity of the problem. Once you have a shorter list of variables, classify them as independent variables, dependent variables, or neither.

Step 3. Define all variables.

It is critical to define all your variables and provide the mathematical notation to be used for each.

Step 4. Select the modeling approach and formulate the model.

Using the tools in this text and your own creativity build a model that describes the situation and whose solution helps to answer important questions.

Step 5. Solve and interpret the model.

We take the model we constructed in Steps 1–4 and solve it. Often, this model might be too complex or unwieldy, so we cannot solve it or interpret it. If this happens, we return to Steps 2–4 and simplify the model further.

Step 6. Verify the model.

Before we use the model, we should test it out. There are several questions we must ask. Does the model directly answer the question, or does the model allow for the answer to the questions to be answered? Is the model usable in a practical sense (can we obtain data to use the model)? Does the model pass the common sense test?

We like to say that we corroborate the reasonableness of our model rather than verify or validate the model.

Step 7. Strengths and weaknesses.

No model is complete without self-reflection of the modeling process. We need to consider not only what we did right but also what we did that might be suspect, as well as what we could do better. This reflection also helps in refining models.

Step 8. Sensitivity Analysis or Model testing.

A modeler wants to know how the inputs affect the ultimate output of any system. Passing the common sense is essential. I once had a class model of Hooke's law with springs and weights. I asked them all to use their models to see how far the spring would stretch using their weight. They all provided the numerical answers, but none said that the spring would break *under their weight*.

Step 9. Implement and maintain the model for future use.

A model is pointless if we do not use it. The more user-friendly the model is, the more it will be used. Sometimes, the ease of obtaining data for the model can dictate its success or failure. The model must also remain current. Often, this entails updating parameters used in the model.

1.5.1.3 Assumptions

In its simplest form, we

Make assumptions
Do some "math"
Derive and interpret conclusions

We say that one cannot question the math, but can question the assumptions used to get to the model used. Assumptions drive the modeling as well as the analysis. According to Albright, every model is based on some set of assumptions. These can be trivial or more complex depending on what we know or can observe about the problem. It might also be affected by the available data.

That is why we say we can question the assumptions and ensure they are justified.

1.6 Illustrate Examples

We now demonstrate the modeling process that was presented in the previous section. Emphasis is placed on problem identification and choosing appropriate (useable) variables. We do not build the models as these modeling examples are repeated later in the book and the models are completed and discussed there.

Example 1.8: Concessions

Consider a concessions firm handling concessions for a sporting event. The manager needs to know whether to stock up with coffee or Coka-Cola. A local agreement restricts you to only one beverage. You estimate a $1,500 profit from selling Coka-Cola if it is cold and a $5,000 profit from selling Coka-Cola if it is warm. You also estimate a $4,000 profit from selling coffee if it is cold and a $1,000 profit from selling coffee if it is warm. The forecast says there is a 30% of a cold front; otherwise, the weather will be warm. What do you do?

> Problem: Using decision theory, what should the concession dealer do?
>
> Assumptions: The weather report holds true. The forecast for weather and customer desires hold true for the event.

Example 1.9: Computer Company Decision

Acme Computer company manufactures memory chips in lots of 10 chips. From past experience, Acme knows 80% of all lots contain 10% (1 out of 10) defective chips, and 20% of all lots contain 50% (5 out of every 10) defective chips. If a good (i.e., 10% defective) batch of chips is sent to the next stage of production, processing costs of $1,000 are incurred, and if a bad batch (50% defective) is sent on to the next stage of production, processing costs of $4,000 are incurred. Acme also has an alternative reworking a batch prior to forwarding it to production at a cost of $1,000. A reworked batch is sure to be a good batch. Alternatively, for a cost of $100, Acme may test one chip from each batch in an attempt to determine whether the batch is defective.

> Problem: Determine how Acme should proceed to minimize expected cost per batch.
>
> Assumptions: There are probabilities involved so perhaps decision-making under uncertainty would be best. A decision tree would be helpful here. The reader can see Bayes' Theorem, in Chapter 2, for another example.
>
> Understanding the problem: There are two states: G = batch is good and B = batch is bad.

Prior probabilities are given as $p(G) = .80$ and $p(B) = .20$

Acme has an option to inspect each batch. The possible outcomes are: D = defective chip and ND = non-defective chip.

We were given the following conditional probabilities:

$$p(D|G) = .10 \quad p(ND|G) = .90 \quad p(D|B) = .50 \quad p(ND|B) = .50$$

Example 1.10: Memory Chips for CPUs

Suppose a small business wants to know how many of two types of high-speed computer chips to manufacture weekly to maximize their profits.

Problem: How many chips of each type are to be produced to maximize profits?

Assumptions: We know the costs and coefficients with certainty. We assume linear programming should work.

Formulate the problem:

First, we need to define our decision variables. Let,

x_1 = number of high speed chip type A to produce weekly
x_2 = number of high speed chip type B to produce weekly

The company reports a profit of $140 for each type A chip and $120 for each type B chip sold. The production line reports the following information:

	Chip A	Chip B	Quantity Available
Assembly time (hours)	2	4	1,400
Installation time (hours)	4	3	1,500
Profit (per unit)	140	120	

The constraint information from the table becomes inequalities that are written mathematically as:

$2x_1 + 4x_2 \leq 1,400$ (assembly time)
$4x_1 + 3x_2 \leq 1,500$ (installation time)
$x_1 \geq 0, x_2 \geq 0$

The profit equation is:

Profit $Z = 140x_1 + 120x_2$

Example 1.11: Batter-Pitcher Dual

Scenario: Consider an upcoming game between the New York Yankees (NYY) and the Los Anglos Dodgers (LAD). Garrett Cole is pitching for the NYY and Shohei Ohtani is batting for the LAD. How should Cole pitcher to Ohtani and how should Ohtani guess about Cole's pitches.

We have the following historical information about the two from numerous previous situations where they faced each other before.

| | | Cole Pitch | | |
		Fastball	Change up	Knuckle Curve
Ohtani	Fastball	.450	.200	.150
Guess	Change up	.284	.330	.195
	Knuckle Curve	.199	.185	.225

What should each player do in this situation? We will apply game theory.

1.7 Technology

In mathematical modeling that we have done, it is impossible to proceed through all the steps without technology. That is why this chapter is called perfect partners. The partnering of technology with modeling is both key and essential to good modeling principles and practices. In this book, we illustrate three different technologies: Excel, Maple, and MATLAB.

1.7.1 Excel

Although Excel might not be the go-to technology for mathematicians or academicians, it is a "go-to" tool for the real world. The more we can empower students in math, science, and engineering to use Excel properly then the better solutions will be for solving future problems.

1.7.2 MAPLE

MAPLE is an excellent technology for mathematics and operations research majors. Its power and graphical interface in two and three dimensions make it an excellent tool.

1.7.3 LINDO

LINDO is linear as well as integer and mixed integer programming software that may be downloaded free from www.LINDO.com.

1.8 Summary

1.8.1 Decision-Making Process

Decision-making is the study of identifying and choosing alternatives based on the values and preferences of the decision-maker. Making a decision implies that there are alternative choices to be considered, and in such a case we want not only to identify as many of these alternatives as possible but also to choose the one that best fits with our goals, objectives, desires, values, and so on. Decision-making should start with the identification of the decision-maker(s) and stakeholder(s) in the decision, reducing the possible disagreement about problem definition, requirements, goals, and criteria. Then, a general decision-making process can be divided into the following steps:

1.8.1.1 Step 1. Define the Problem

This process must, as a minimum, identify root causes, limiting assumptions, system and organizational boundaries and interfaces, and any stakeholder issues.

The goal is to express the issue in a clear, one-sentence problem statement that describes both the initial conditions and the desired conditions. Of course, the one-sentence limit is often exceeded in the practice in case of complex decision problems. The problem statement must however be a concise and unambiguous written material agreed upon by all decision-makers and stakeholders. Even if it can sometimes be a long iterative process to come to such an agreement, it is a crucial and necessary point before proceeding to the next step.

1.8.1.2 Step 2. Determine Requirements

Requirements are conditions that any acceptable solution to the problem must meet. Requirements spell out what the solution to the problem must do. In mathematical form, these requirements are the constraints describing the set of the feasible (admissible) solutions of the decision problem. It is very important that even if subjective or judgmental evaluations may occur in the following steps, the requirements must be stated in exact quantitative form, i.e., for any possible solution, it has to be decided unambiguously whether it meets the requirements or not. We can prevent the ensuing debates by putting down the requirements and how to check them in a written material.

1.8.1.3 Step 3. Establish Goals

Goals are broad statements of intent and desirable programmatic values. ... Goals go beyond the minimum essential must have's (i.e., requirements) to wants and desires. In mathematical form, the goals are objectives contrary to the requirements that are constraints. The goals may be conflicting, but this is a natural concomitant of practical decision-making situations.

1.8.1.4 Step 4. Identify Alternatives

Alternatives offer different approaches for changing the initial condition into the desired condition. Be it an existing one or only constructed in mind, any alternative must meet the requirements. If the number of the possible alternatives is finite, we can check one by one if it meets the requirements. The infeasible ones must be deleted (screened out) from the further consideration, and we obtain the explicit list of the alternatives. If the number of the possible alternatives is infinite, the set of alternatives is considered as the set of the solutions fulfilling the constraints in the mathematical form of the requirements.

1.8.1.5 Step 5. Define Criteria

Decision criteria, which will discriminate among alternatives, must be based on the goals. It is necessary to define discriminating criteria as objective measures of the goals to measure how well each alternative achieves the goals. Since the goals will be represented in the form of criteria, every goal must generate at least one criterion, but complex goals may be represented only by several criteria.

We suggest that when we have multiple criteria that we attempt to prioritize them in their order of importance.

Consider this when coming up with criteria: criteria should be

- able to discriminate among the alternatives and to support the comparison of the performance of the alternatives,
- complete to include all goals,
- operational and meaningful,
- non-redundant,
- few in number.

1.8.1.6 Step 6. Select a Decision-Making Tool

There are several tools for solving a decision problem. We will learn many during this course. The selection of an appropriate tool is not an easy task and depends on the concrete decision problem as well as on the objectives of the decision-makers.

Sometimes, the simpler the method, the better, but complex decision problems may require complex methods as well.

1.8.1.7 Step 7. Evaluate Alternatives against Criteria

Every correct method for decision-making needs, as input data, the evaluation of the alternatives against the criteria. Depending on the criterion, the assessment may be objective (factual), with respect to some commonly shared and understood scale of measurement (e.g., money) or can be subjective (judgmental), reflecting the subjective assessment of the evaluator. After the evaluations, the selected decision-making tool can be applied to rank the alternatives or to choose a subset of the most promising alternatives.

1.8.1.8 *Step 8. Validate Solutions against Problem Statement*

The alternatives selected by the applied decision-making tools have always to be validated against the requirements and goals of the decision problem. It may happen that the decision-making tool was misapplied. In complex problems, the selected alternatives may also call the attention of the decision-makers and stakeholders that further goals or requirements should be added to the decision model.

Exercises

1.1 How would you approach a problem concerning a drug dosage? Do you always assume the doctor is right?

1.2 In modeling the size of any prehistoric creature, what information would you like to be able to obtain? What additional assumptions might be required?

1.3 In the oil rig problem, what other factors might be critical in obtaining a "good" model that predicts reasonably well? What variables could be important that were not considered?

1.4 For the model in Example 1.3, are the assumptions about the data reasonable? How would you collect data to build the model? What other variables would you consider?

1.5 In the bridge example, what is the impact of the location of the bridge?

1.6 In the bank queue problem, discuss the criticality of the assumptions. Do you feel that more training is as valuable as adding another server?

1.7 In Example 1.8, review your algebra and trigonometry and construct the region to be searched. Describe the shape and find the overall area to be searched.

Projects

1.1 Is Michael Jordan the greatest basketball player of the century? What variables and factors need to be considered?

1.2 What kind of car should you buy when you graduate from college? What factors should be in your decision? Are car companies modeling your needs?

1.3 Consider domestic decaffeinated coffee brewing. Suggest some objectives that could be used if you wanted to market your new brew. What variables and data would be useful?

1.4 Replacing a coaching legend at a school is a difficult task. How would you model this? What factors and data would you consider? Would you equally weigh all factors?

1.5 How would you go about building a model for the "best pro football player of all time"?

1.6 Rumors abound in major league baseball about steroid use. How would you go about creating a model that could imply the use of steroids? Relate baseball's steroid rules to the Yankee's Alex Rodriquez case.

1.7 After the 2009 All Star baseball game, the America League has won 12 straight games dating back to 1997. This is the longest winning straight by either side. Help the National League prepare to win by designing a model for players or a line-up that could help them change their outcome.

References

Albright, B. (2010). *Mathematical Modeling with Excel*. Sudbury, MA: Jones and Bartlett, 2010.

Giordano, F., W. Fox, & S. Horton. (2014). *A First Course in Mathematical Modeling*, 5th ed. Boston, MA: Cengage Publishers.

Additional Readings

Albright, B. & W. Fox. (2020). *Mathematical Modeling with Excel*, 2nd ed. Boca Raton, FL: Taylor and Francis Publishers.

Burden, R. & D. Faires. (1997). *Numerical Analysis*. Pacific Grove, CA: Brooks-Cole.

COMAP. Modeling Competition Sites found at www.comap.com/contests

Fox, W. (2018). *Mathematical Modeling for Business Analytics*. Boca Raton, FL: Taylor and Francis Publishers.

2

Decision Theory and Expected Value

2.1 Introduction

Decision theory in mathematics and statistics is concerned with identifying the values, uncertainties, and other issues relevant in a given decision and the resulting optimal decision.

The environment within decisions theory where decisions are made is often categorized into four states: certainty, risk, uncertainty, and conflict. Decision theory, per se, is primarily concerned with decisions under the conditions of risk and of uncertainty. Making decisions under certainty has been studied before and will be reviewed and uses linear programming. Linear programming is essential and it is also a method of solving game theory problems. Game theory is also called games (decisions) concerning conflict.

The state of *certainty* assumes when all the information required to make a decision is assumed known and true. The condition of *risk* exists when perfect information is not available but the probabilities that certain outcomes will occur can be estimated. A state of uncertainty exists when the probabilities of occurrence in a decision situation are not known. Certainty and uncertainty are the two extremes and risk is someone in between. Conflict exists when two or more decision-makers are in competition.

Hence, decision-makers are not only interested in their decisions but also in the other decision-makers.

Choices under Uncertainty: This area represents the heart of decision theory. The procedure now referred to as **expected value** was known from the 17th century. Blaise Pascal invoked it in his famous wager published in 1670. The idea of expected value is that, when faced with a number of actions, each of which could give rise to more than one possible outcome with different probabilities, the rational procedure is to identify all possible outcomes, determine their values (positive or negative) and the probabilities that will result from each course of action, and multiply the two to give an expected value. The action to be chosen should be the one that gives rise to the highest total expected value. In 1738, Daniel Bernoulli published an influential paper entitled *Exposition of a New Theory on the Measurement of Risk*. He also gives

DOI: 10.1201/9781032726885-2

an example in which a Dutch merchant is trying to decide whether to insure a cargo being sent from Amsterdam to St Petersburg in winter, when it is known that there is a 5% chance that the ship and cargo will be lost. In his solution, he defines a utility function and computes expected utility rather than expected financial value.

Since **expected value** is so important to our next lesson on decision theory, so we will spend some time on it in this first section.

2.2 Expected Value

Expected Value is taught in all undergraduate probability and statistics courses. If necessary, you might also want to review that material.

Notation: *E[X]*

Meaning: Expected value is the *mean* or *average value*. There are lots of ways to calculate the average value. We present a few common methods that you could use in decision theory.

Calculating *E[X]*

Average: Two scores 80 and 100. (80+100)/2 = 90

Let's assume you have the following numerical grades in a course: 100, 90, 80, 95, 100. The average or mean, $E[X] = \frac{\sum_{i=1}^{n} x_i}{n} = \frac{100+90+80+95+100}{5} = \frac{465}{5} = 93$

Weighted average: Let's assume in a class there were 8 scores of 100, 5 scores of 95, 3 scores of 90, 2 scores of 80, and 1 score of 75. Find the average?

$$E[X] = \frac{\sum w_i x_i}{\sum w_i} = \frac{(8*100+5*95+2*80+1*75)}{8+5+3+2+1} = \frac{1,780}{19} = 93.68$$

Probabilistic mean: Finding the expected value when probabilities are involved: Number of attempted shop-lifting incident in Harris Teeter each week.

X	1	2	3	4
Probability, $P(X = x)$	.5	.33	.10	.07

$$E[X] = \sum X \cdot P(X = x),$$

where X is a random variable. Recall a random variable is a rule that assigns a number to every outcome in a sample space.

A **random variable** is a numerical measure of the outcome from a probability experiment, so its value is determined by chance. Random variables are denoted using capital letters such as X. The values that they take on are annotated by small letters, x. We would say the $P(X = x)$, probability that the random variable X takes on a values x.

$$E[X] = (1)*(.5)+(2)*(.33)+(3)*(.10)+(4)*(0.07) = 1.74$$

There are, on average, 1.74 shop-lifters per week in Harris Teeter.

Another definition: Expected value is $p_i o_i$, where p is a probability and o represents an outcome.

Example 2.1: Expected Value

If the $P(s) = 1/5$ and if successful we make \$50,000 and if unsuccessful we lose \$1,000, find the expected value.

Solution:

$$50,000(1/5)-1,000(4/5) = 10,000-800 = \$9,200$$

Example 2.2: Life Insurance Expected Value

A term life insurance policy will pay a beneficiary a certain sum of money upon the death of the policy holder. These policies have premiums that must be paid annually. Suppose a life insurance company sells a \$250,000 one year term life insurance policy to a 49-year-old female for \$550. According to the National Vital Statistics Report, Vol. 47, No. 28, the probability the female will survive the year is 0.99791. Compute the expected value of this policy to the insurance company.

Solution and Decision:

$$E[X] = 550*(1)-250,000*(1-.99791) = 27.5$$

Note that since the $E[X] > 0$, then we should sell the policy. If $E[X] < 0$, then we should not sell the policy.

Example 2.3: Gambling

A common application of expected value is in gambling. For example, an American roulette wheel has 38 equally likely outcomes. A winning bet placed on a single number pays 35-to-1 (this means that you are paid 35 times your bet and your bet is returned, so you get 36 times your bet). So considering all 38 possible outcomes, the expected value of the profit resulting from a \$1 bet on a single number. The *house average* or *house edge* (also called the <u>expected value</u>) is the amount the player loses relative for any bet made, on average. If a player bets on a single number in the American game, there is a probability of 1/38 that the player wins 35 times the bet, and a 37/38 chance that the player loses their bet. The expected value is:

$$-1 \times (37/38) + 35 \times (1/38) = -0.0526 \ (5.26\% \text{ house edge}).$$

This is about −$0.0526. Therefore, one expects, on average, to lose over five cents for every dollar bet, and the **expected value** of a one dollar bet is $0.9474. In gambling or betting, a game or situation in which the expected value of the profit for the player is zero (no net gain nor loss) is commonly called a "fair game".

$$\left(-\$1\ (37/38)\right)+(1/38) = \$0.9474$$

The Average Expected Payoff: An estimate of the amount that will be gained in a game of chance, calculated by multiplying the probability of winning by the number of points won each time.

2.3 Decisions under Risk: Probabilities are Known or Estimated in Advance

Decisions under risk uses Expected Value or Expected Opportunity Loss to compare decisions:

2.3.1 Expected Value (Realist)

Compute the expected value under each action and then pick the action with the largest expected value. This is the only method of the four that incorporates the probabilities of the states of nature. The expected value criterion is also called the Bayesian principle.

2.3.2 Minimax (Opportunist)

Minimax decision-making is based on opportunistic loss. They are the kind that looks back after the state of nature has occurred and say "Now that I know what happened, if I had only picked this other action instead of the one I actually did, I could have done better". So, to make their decision (before the event occurs), they create an opportunistic loss (or regret) table. Then they take the **min**imum of the **max**imum. That sounds backward, but remember, this is a *loss* table. This similar to the maximin principle in theory; they want the best of the worst losses.

Example 2.4: Concessions

Consider a concessions firm handling concessions for a sporting event. The manager needs to know whether to stock up with coffee or coke-cola. A local agreement restricts you to only one beverage. You estimate a $1,500 profit selling coke-cola if it is cold and a $5,000 profit of selling cola if it is warm. You also estimate a $4,000 profit selling coffee if it is cold and a $1,000 profit of selling coffee if it is warm. The forecast says there is a 30% of a cold front otherwise the weather will be warm. What do you do?

Decisions under risk (probabilities known) use Expected value.

$$E[cola] = 1,500*.3 + 5,000*.7$$
$$= \$3,950 \; E[coffee] = 4,000*.3 + 1,000*.7 = \$1,900$$

$$E[cola] > E[coffee]$$

Decision is to sell cola

What if analysis?

Under what conditions should we sell coffee?

Assume $p1$ = probability the weather is warm and $p2$ = probability the weather is cold where $p1 + p2 = 1$.

Set the two Expected values equal and solve.

$$E[cola] = E[coffee] \; 1,500*p2 + 5,000*p1 = 4,000*p2 + 1,000*p1$$

Let $p1 = 1 - p2$ and substitute.

$$1,500 \; p2 + 5,000(1-p2) = 4,000 \; p2 + 1,000(1-p2)$$

$$1,500 \; p2 + 5,000 - 5,000 \; p2 = 4,000 \; p2 + 1,000 - 1,000 \; p2$$

$$4,000 = 6,500 \; p2$$

$$p2 = 4,000/6,500 = 0.615 \; (\text{to 3 decimal places})$$

When the probability of cold weather is greater than 0.615, we sell coffee. Otherwise sell cola.

Example 2.5: Expected Opportunity Loss

This is an alternate criterion under risk. The object is to minimize the expected regret experienced as a result of a decision. Consider the following:

Alternatives	Nature	
Investments	Condition 1, $p = 0.4$	Condition 2 $q = (1 - p) = 0.6$
A	$50,000	−$10,000
B	$15,000	$60,000
C	$100,000	$10,000

We create an Opportunity Loss table. We consider the best we can do in each State of nature as the base. We calculate the difference of each choice from the base. We look at each state of nature and pick the largest value and then subtract each alternative outcome form that value. For each, in state of nature 1, the best is 100,000 and we subtract each alternatives value from 100,000 to get the new first column. We do for column 2 as well to obtain:

Alternatives	Nature	
Investments	Condition 1, $p = 0.4$	Condition 2, $q = 0.6$
A	$50,000	$70,000
B	$85,000	$0
C	$0	$50,000

$$EOL[A] = (.4)*50,000 + .6*70,000 = 62,000$$

$$EOL[B] = (.4)*85,000 + .6*0 = 34,000$$

$$EOL[C] = .4*0 + .6*50,000 = 30,000$$

Since *EOL[C]* is the smallest, choose alternative C because it minimizes regret.

2.4 Decisions under Uncertainty: Probabilities are Not Known nor Can They be Estimated

There are five possible criteria all of which may lead to different decisions. You must decide on the criterion first based on your knowledge of your decision-maker.

Five decision criteria under risk are:

Laplace criterion

Maximin criterion

Maximax criterion

Hurwicz criterion, and

Regret criterion

Laplace assumes equally likely probabilities. **Maximin** maximizes the minimum return.

Maximax maximizes the maximum payoff. **Hurwicz** is a compromise of the *Maximin* and *Maximax criterion*. **Regret** is based on opportunity loss where we minimize loss.

Example 2.6: Applying the Criteria

Assume we have 100,000 to invest (one time). We have three alternatives: stocks, bonds, or savings. We place the estimated returns in a payoff matrix.

Alternatives	Conditions		
Investments	Fast Growth (Risk)	Normal Growth	Slow Growth
Stocks	$10,000	$6,500	−$4,000
Bonds	$8,000	$6,000	$1,000
Savings	45,000	$5,000	$5,000

<u>Laplace criterion</u>: This assumes equal probabilities. Since there are three conditions, the chance of each is 1/3. We compute the expected value.

$$E[Stocks] = 1/3(10,000) + 1/3(6,500) + (1/3)(-4,000) = 4,167$$

$$E[Bonds] = 1/3\ (8,000) + 1/3\ (6,000) + 1/3\ (1,000) = 5,000$$

$$E[Savings] = 1/3(5,000) + 1/3\ (5,000) + 1/3\ (5,000) = 5,000$$

We have a tie so we can do either bonds or savings under the Laplace criterion.

<u>Maximin criterion</u>: This assumes the decision-maker is pessimistic about the future. According to this criterion, the minimum return for each alternative are compared and we choose the maximum of these minimums.

Stocks −4,000
Bonds 1,000
Savings 5,000

The maximum of these is Savings at 5,000.

<u>Maximax criterion</u>: This assumes an optimistic decision-maker. We take the maximum of each alternative and then the maximum of those.

Stocks 10,000
Bonds 8,000
Savings 5,000

The maximum of these is 10,000 with Stocks.

Hurwicz criterion: This is a compromise of the maximin and the maximax. It is a mixture of pessimism and optimism. It requires a coefficient of optimism, called α that is a real value that is between 0 and 1.

Assume the coefficient of optimism is 0.6.

	Maximin	Maximax
Stocks	−4,500	10,000
Bonds	1,000	8,000
Savings	5,000	5,000

We compute the expected values using:

$$\alpha * \text{Optimism value} + (1-\alpha) * \text{Pessimism value}$$

We illustrate this as follows:

$$EH[Stocks] = .6(10,000) + .4(-4,500) = 4,400$$

$$EH[Bonds] = .6(8,000) + .4(1,000) = 5,200$$

$$EH[Savings] = .6(5,000) + .4(5,000) = 5,000$$

5,200 is the best answer. So with our choice of α (as 0.6), we pick Bonds. We note that the decisions might change based on your choice of α.

Regret Criterion: The underlying principle is that the decision-maker experience regret when a state of nature occurs that causes the selected alternative to realize less than the maximum payoff. We build a table of opportunity loss. We take the base as the maximum in each state of nature. The following is our regret matrix for this example.

Alternatives	Conditions		
Investments	Fast Growth (Risk)	Normal Growth	Slow Growth
Stocks	$10,000 − 10,000 = 0	6,500 − $6,500 = 0	5,000 − (−$4,000) = 9,000
Bonds	10,000 − 8,000 = 2,000	6,500 − $6,000 = 500	5,000 − 1,000 = 4,000
Savings	10,000 − 5,000 = 5,000	6,500 − $5,000 = 1,500	5,000 − 5,000 = 0

Now, make a sub-table of Maximum regrets.

Stocks 9,000

Bonds 4,000

Savings 5,000

Choose the least regret, which is Bonds.
Summary of our decisions:

Criterion	Choice
Laplace	Saving or Bonds
Maximin	Savings
Maximax	Stocks
Hurwicz	Bonds (with $\alpha = 0.6$)
Regret	Bonds

Next the characteristics of the decision-maker must be taken into account before you choose the criterion option.

2.4.1 Some Definitions

Expected Value (Realist)

Compute the expected value under each action and then pick the action with the largest expected value. This is the only method of the four that incorporates the probabilities of the states of nature. The expected value criterion is also called the Bayesian principle.

Maximax (Optimist)

The maximax looks at the best that could happen under each action and then chooses the action with the largest value. They assume that they will get the most possible and then they take the action with the best "best case" scenario. The **maxim**um of the **max**imums or the "best of the best". This is the lotto player; they see large payoffs and ignore the probabilities.

Maximin (Pessimist)

The maximin person looks at the worst that could happen under each action and then choose the action with the largest payoff. They assume that the worst that can happen will, and then they take the action with the best "worst case" scenario. The **maxim**um of the **min**imums or the "best of the worst". This is the person who puts their money into a savings account because they could lose money in the stock market.

Minimax (Opportunist)

Minimax decision-making is based on opportunistic loss. They are the kind that look back after the state of nature has occurred and say "Now that I know what happened, if I had only picked this other action instead of the one I actually did, I could have done better". So, to make their decision (before the event occurs), they create an opportunistic loss (or regret) table. Then they take the **min**imum of the **max**imum. That sounds backward, but remember, this is a *loss* table. This similar to the maximin principle in theory; they want the best of the worst losses.

2.5 Decision Trees

2.5.1 Decision Trees and Bayes Theorem

Often we must make a series of decisions at different points in time. Here decision trees can be used. A decision tree allows us to decompose large complex decision problems in several smaller problems.

Expected value of sample information (EVWSI)

Expected value of perfect Information (EVWPI)

A **decision tree** is a decision support tool that uses a tree-like graph or model of decisions and their possible consequences, including chance event outcomes, resource costs, and utility. It is one way to display an algorithm. Decision trees are commonly used in operations research, specifically in decision analysis, to help identify a strategy most likely to reach a goal. Another use of decision trees is as a descriptive means for calculating.

A decision tree consists of three types of nodes:

1. Decision nodes – commonly represented by squares
2. Chance nodes – commonly represented by circles
3. End nodes – commonly represented by triangles

Example 2.7: US Army Recruiting Command

The recruiting command has assets of $1.5 million and wants to decide whether or not to market a new strategy. They have three alternatives.

Alternative 1. Test the new strategy locally, then they utilize results of the market study to determine whether or not to market nationally.

Alternative 2. Without any test markets, market the new strategy nationally.

Alternative 3. Without testing decide not to market the new strategy.

Facts: In the absence of a market study, the command believes that the new strategy has a 55% chance of being a national success and a 45% chance of being a national failure. If it is a national success, the assets will increase to $3 million and if a failure will lose $1 million.

If the command performs a market study (at a cost of $300,000), there is a 60% chance that the study will yield a successful outcome (a local success) and a 40% chance of an unsuccessful result (a local failure). If a local success is observed, there is an 85% chance of national success. If a local failure, then there is only a 10% chance of national success. If the command wants to maximize its expected value, what strategy should the command follow?

This is depicted in Figure 2.1. We build the tree from right to left and fill in the unknowns from left to right. We will do this in class to reduce the mystery. The result is an expected value of 2.7 million by not doing a test market and marketing nationally.

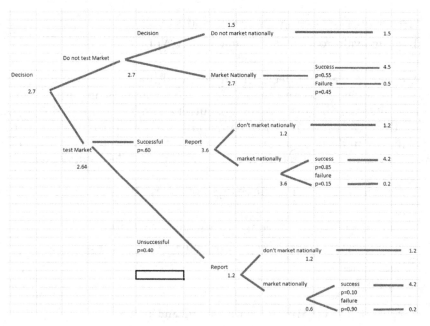

FIGURE 2.1
Recruiting command example.

Example 2.8: Art Dealer Decisions

An art dealer is willing to purchase a painting for $50,000. The dealer can buy the painting today for $40,000 or can wait and buy it tomorrow (if it has not already sold) for $30,000. The dealer might wait another day (if still unsold) and buy the $26,000. At the end of the 3rd day, the painting will no longer be available. Each day, he estimates that there is probability of 60% that the painting will be sold. What decision strategy should the painter take?

Stage 1: buy $10,000

Not buy Stage 2 Sold ($p = 0.6$) $0

Has not sold ($p = 0.4$) buy $20,000

Don't buy sold ($p = 0.6$) $0

Not sold ($p = 0.4$) buy| $24,000 (left)

Don't buy: $0

We should buy immediately.

2.6 Sequential Decisions and Conditional Probability*

In many cases, decisions must be made sequentially. Often decisions are based on other or previous made decisions. These are multi-stage decisions. We will examine the method and effects of sequential decisions. Often, a key part to decisions is to compute or use conditional probabilities.

Example 2.9: Hardware & Lumber Company Decision with Multi-Stage Choices

Consider a company that needs to decide whether or not to build and market outdoor play sets. The three alternatives under consideration with the respective demand revenues and loses with the estimated demand probabilities are as follows:

	Outcomes		
	High Demand	**Moderate Demand**	**Low Demand**
Alternatives	($p = 0.35$)	($p = 0.40$)	($p = 0.25$)
Large plant	$200,000	$120,000	−$120,00
Small plant	$90,000	$55,000	−$20,000
No plant	$0	$0	$0

Initial analysis shows we build the large plant.

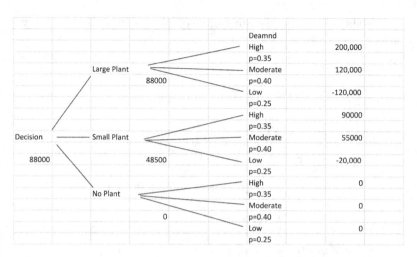

Let's assume that before making a decision, the company has an option to hire a market research company for $4,000. This company will survey the markets that are serviced by this company as to the attractiveness of the new outdoor play sets. The company knows that the market research does not provide perfect information but does provide updated information based on their sample survey. In addition, the company has to decide whether or not to hire the market research team. If the research is conducted, the assumed probabilities of a success survey is 0.57 and an unsuccessful survey is 0.43. Further, since we have gained more information, our probabilities for the demand will change. Given a successful survey outcome, the probability of high demand is 0.509, for moderate demand is 0.468, and/or for low demand is 0.023. Given an unsuccessful survey outcome, the probability of high demand is 0.023, for moderate demand is 0.543, and/or for low demand is 0.434.

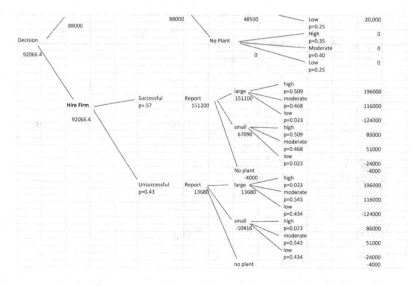

Decision: We conclude that we should hire and conduct the market survey to obtain the expected value of $92,066.40.

Example 2.10: Computer Company Decision Tree

Acme Computer company manufactures memory chips in lots of 10 chips. From past experience, Acme knows 80% of all lots contain 10% (1 out of 10) defective chips, and 20% of all lots contain 50% (5 out of every 10) defective chips. If a good (that is 10% defective) batch of chips is sent to the next stage of production, processing costs of $1,000 are incurred, and if a bad batch (50% defective) is sent on to the next stage of production, processing cost of $4,000 are incurred. Acme also has an alternative reworking a batch prior to forwarding to production at a cost of $1,000. A reworked batch is sure to be a good batch. Alternatively, for a cost of $100, Acme may test one chip from each batch in an attempt to determine whether the batch is defective. Determine how Acme should proceed to minimize expected cost per batch.

Solution:

There are two states: G = batch is good and B = batch is bad.

   Prior probabilities are given as $p(G) = .80$ and $p(B) = .20$

   Acme has an option at inspect each batch. The possible outcomes are: D = defective chip and ND = non-defective chip.

Probability:

We introduce three new probability concepts: independence, conditional probability, and Bayes' theorem. Independence between events occurs when the chance of one event happening does not affect the chance of happening of other events. Mathematically, we state that if we have two events A and B, that events A and B are independent if and only if $p(A \cap B) = p(A) * p(B)$. The second concept condition probability allows the knowledge of one event to possibly influence the probability of a second event. Mathematically, we state $p(A|B)$, the probability of event A given event B has occurred, as follows in equation (1):

$$p(A|B) = \frac{p(A \cap B)}{p(B)} \qquad (1)$$

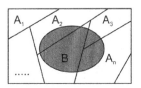

FIGURE 2.2
Venn diagram of events B and A_i.

Bayes' theorem is illustrated with a simple Venn diagram (Figure 2.2). We want to know the probability of event B but all we know are the intersections of B with event A. By summing all the known intersections, we are able to determine the probability of event B. With these new concepts, we proceed with our example. The resulting formula is given in equation (2).

We were given the following conditional probabilities:

$$p(D\,|\,G) = .10 \quad p(ND\,|\,G) = .90 \quad p(D\,|\,B) = .50 \quad p(ND\,|\,B) = .50$$

To complete and find our values required, we need to determine some additional probabilities. We will use two sets of formulas: conditional probability (equation 2) and Bayes' theorem (equation 2).

$$p(a\,|\,b) = \frac{p(a \cap b)}{p(b)} \tag{2}$$

$$p(a_j\,|\,b) = \frac{p(a_j \cap b)}{p(b)} \tag{3}$$

We start with a tree for Bayes' theorem initially as in Figure 2.3.

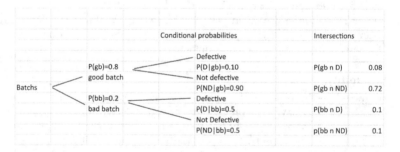

FIGURE 2.3
Bayes' theorem tree diagram.

$$p(D) = p(D \cap gb) + p(D \cap bb) = .10 + .08 = .18$$

$$p(ND) = p(ND \cap gb) + p(ND \cap bb) = .72 + .10 = .82$$

We can now find our needed probabilities for our decision model.

$$p(bb\,|\,D) = \frac{p(bb \cap D)}{p(D)} = \frac{.10}{.18} = 0.55555$$

$$p(gb\,|\,D) = 1 - .5555555 = .45555555$$

$$p(bb\,|\,ND) = \frac{p(bb \cap ND)}{p(ND)} = \frac{.10}{.82} = 0.12195$$

$$p(gb\,|\,ND) = 1 - .12195 = .87805$$

Decision for computer company: Our decision is to test the chip as it has the minimum expected cost of $1,580 as compared to $1,600 to send the batch on as is or compared to $2,000 to rework the batch.

Decision tree probabilities can be based on a manager's experience and intuition, historical data, or computed from other data using Bayes' theorem:

> the Bayes' theorem approach recognizes that a decision-maker does not know with certainty which state of nature will occur
> initial (prior) probability assessments can be updated with additional information as it becomes available
> updated probability assessments are called revised or posterior probabilities

Probability example:

A cup contains two dice that are identical in appearance. One is fair (unbiased), while the other is loaded (biased). The probability of rolling a 3 on the fair die is 1/6 while the probability of a 3 on the loaded die is 0.60.

Not knowing which die is which, one is selected at random and tossed. The result is a 3. Given this additional information, what is the revised probability that the die was fair?

1. Take stock of the information and probabilities available (prior probabilities)
   since the die was selected randomly from the 2 dice in the cup, the probability of it being fair or loaded is 0.5

$$P(\text{fair}) = 0.50 \quad P(\text{loaded}) = 0.50$$

   the probability of rolling a 3 on the fair die is 1/6, while the probability of a 3 on the loaded die is 0.60

$$P(3 \,|\, \text{fair}) = 0.167 \quad P(3 \,|\, \text{loaded}) = 0.600$$

2. Compute joint probabilities for P(3 and fair), P(3 and loaded)

$$P(AB) = P(A \,|\, B) \times P(B)$$
$$P(3 \text{ and fair}) = P(3 \,|\, \text{fair}) \times P(\text{fair})$$
$$= (0.167)(0.50)$$
$$= 0.083$$
$$P(3 \text{ and loaded}) = P(3 \,|\, \text{loaded}) \times P(\text{loaded})$$
$$= (0.600)(0.50)$$
$$= 0.300$$

Example 2.11: Rolling Die

  a. The unconditional probability of rolling a 3 with a fair or loaded die is 0.083 + 0.3 = 0.383
  b. The revised probability given a 3 was rolled:
     P(fair die | 3 was rolled) = P(fair die ∩ 3 was rolled)/P(3 was rolled) = 0.083/.383 = 0.22
     P(loaded die | 3 was rolled) = P(loaded die ∩ 3 was rolled)/P (3 was rolled) = 0.3/.383 = 0.78

- Before the die was rolled, the best probability assessment available was that there was a 50-50 chance that is fair and a 50-50 chance that it was loaded
- One roll of the die provided additional information that allowed the prior probability to be revised to a posterior probability
  the new posterior estimate is that there is a 0.78 probability that the die rolled was loaded and only a 0.22 probability that it was fair.

Now that we have more information on Bayes' Rule, let's continue.

Thompson Lumber

It was previously assumed that the following conditional probabilities were known:

P(favorable market | favorable survey) = P(FM | FS) = 0.78

P(unfavorable market | favorable survey) = P(UM | FS) = 0.22

P(favorable market | unfavorable survey) = P(FM | US) = 0.27

P(unfavorable market | unfavorable survey) = P(UM | US) = 0.73

- These values can be derived using Bayes' theorem.
- From discussions with market research specialists, it is known that special surveys can either be positive (predicting a favorable market) or negative (predicting an unfavorable market).
- It is also known that for recent favorable markets, market surveys were accurate and correctly predicted a favorable market 70% of the time. However, the market surveys were inaccurate and incorrectly predicted an unfavorable market 30% of the time.

For recent unfavorable markets, market surveys were accurate and correctly predicted an unfavorable market 80% of the time. However, this leaves 20% of the time when the surveys were inaccurate and incorrectly predicted a favorable market.

We need to compute some conditional probabilities shown in Figure 2.4:

Thompson Lumber Co.

$$P(UM \mid FS) = \frac{P(FS \mid UM)P(UM)}{P(FS \mid UM)P(UM) + P(FS \mid FM)P(FM)}$$

$$= \frac{(0.20)(0.50)}{(0.20)(0.50) + (0.70)(0.50)} = \frac{0.10}{0.45} = 0.22$$

$$P(FM \mid US) = \frac{P(US \mid FM)P(FM)}{P(US \mid FM)P(FM) + P(US \mid UM)P(UM)}$$

$$= \frac{(0.30)(0.50)}{(0.30)(0.50) + (0.80)(0.50)} = \frac{0.15}{0.55} = 0.27$$

$$P(UM \mid US) = \frac{P(US \mid UM)P(UM)}{P(US \mid UM)P(UM) + P(US \mid FM)P(FM)}$$

$$= \frac{(0.80)(0.50)}{(0.80)(0.50) + (0.30)(0.50)} = \frac{0.40}{0.55} = 0.73$$

FIGURE 2.4
Thompson lumber with Bayes' theorem applied. (From Operation Research: Applications and Algorithm, 4th Ed by Wayne Winston, Brooks-Cole (Cengage) Publisher, 2004.)

Example 2.12: Colaco Decision

Colaco has $150,000 in assets and wants to know whether or not to market a new product. They have three alternatives:

Alternative 1: Test market the new product and use the results to decide whether to market the new product nationally.
Alternative 2: Immediately market the new product without any testing.
Alternative 3: Immediately decide not to market without any testing.

It is estimated that with a market study that Colaco has a 55% chance of being a national success and a 45% chance of being a national failure. If the new product is a national success, assets increase by $300,000 and if the new product is a national failure, the assets decrease by $100,000.

A market study costs $30,000. It is thought that there is a 60% chance that the market study will be favorable for the new product and a 40% chance that the study will show an unfavorable result. If the market study shows a success, then there is an 85% chance that the new product is a national success. If the market survey yields an unfavorable results, there is only a 10% chance that the new product will be a national success. This is shown in Figure 2.5. What strategy should the company follow to maximize expected value?

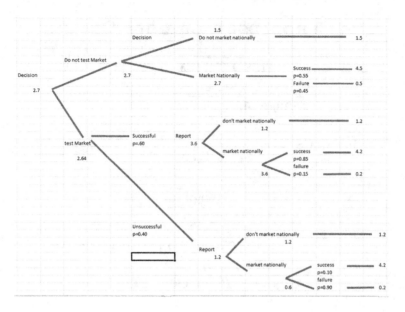

FIGURE 2.5
Colaco decision tree.

Now, let's assume we have more information. The experiment of the test marketing yielded

$$P(NS|LS) = 0.85 \quad P(NS|LF) = 0.10$$

$$P(NF|LS) = 0.15 \quad P(NF|LF) = 0.90$$

Example 2.13: Tech Computers

Tech computers manufacturer memory chips in lots of ten chips. From past experience they know that every 80% of the lots have 10% (1 out of 10) defective chips and the other 20% of the lots contain 50% (5 out of 10) defective chips. If a good batch is sent on (assume that is our 80% good and 10% defective, a processing cost of $1,000 is incurred. If the 20% batch with 50% defective is sent forward than a cost of $4,000 is incurred. They also have an alternative of reworking an entire batch for a cost of $1,000. A reworked batch is assured of being 100%. Also, they could test one chip from a batch for $100 to determine whether is comes from the 10% or 50% defective lot. Tech wants to minimize the expected cost per batch.

G = good batch
B = bad batch

We know $P(G) = 0.80$ and $P(B) = 0.20$.
Tech can inspect a chip to determine if defective (D) or not defective (ND). We have some associated conditional probabilities.

$$P(D|G) = 0.10 \; P(ND|G) = 0.90 \; P(D|B) = 0.50 \; P(ND|B) = 0.50$$

In order to complete the decision tree diagram, we need to compute some additional probabilities:

P(D n G) = 0.08
P(D n N) = 0.10
P(ND n G) = 0.72
P(ND n B) = 01.0

From these, we use the law of total probability to determine:

$$P(D) = P(D \text{ n } G) + P(D \text{ n } B) = 0.08 + 0.10 = 0.18$$

$$P(ND) = P(ND \text{ n } G) + p(ND \text{ n } B) = 0.72 + 0.10 = 0.82$$

With this we can compute the required posterior probabilities:

P(B|D) = 5/9
P(G|D) = 4/9
P(B|ND) = 10/82
P(G|ND) = 72/82

We find we are better off testing the chip and making decisions based on the tested chip. The expected cost is $1,580.

Chapter 2 Exercises

2.1 Let's assume you have the following numerical grades in a course: 100, 90, 80, 95, 100. Compute your average grade.

2.2 Weighted means: Let's assume in a class there were 8 scores of 100, 5 scores of 95, 3 scores of 90, 2 scores of 80, and 1 score of 75. Find the average grade.

2.3 Finding the expected value when probabilities are involved. The number of attempted ATM uses per person and their probabilities are listed below. Compute the expected value and interpret that value.

	1	2	3	4
Probability	.5	.33	.10	.07

2.4 Assume the probability of success is, $P(s) = 1/5$, and further assume that if we are successful that we make $50,000 and if we are unsuccessful we lose $1,000. Find the expected value.

2.5 A term life insurance policy will pay a beneficiary a certain sum of money upon the death of the policy holder. These policies have premiums that must be paid annually. Suppose a life insurance company sells a $250,000 one year term life insurance policy to a 49-year-old female for $550. According to the National Vital Statistics Report, Vol. 47, No. 28, the probability the female will survive the year is 0.99791. Compute the expected value of this policy to the insurance company.

2.6 Consider a construction firm which is deciding to specialize in building high schools or elementary schools, or a combination of both over the long haul. The construction company must submit a bid proposal which costs money to prepare and there are no guarantees that they will be awarded the contract. If they bid on the high school, they have a 25% of getting the contract and they expect to make $62,000 net profit. However, if they do not get the contract, they lose $1,500. If they bid on the elementary school, there is a 20% of getting the contract. For the elementary school, they would net $40,000 in profit. However, if they do not get the contract, they lose $750. What should the construction company do?

2.7 Company ABC has developed a new line of products. Top management is attempting to decide on both marketing and production strategies. Three strategies are considered, and will be referred to as A (aggressive), B (basic), and C (cautious). The conditions under which the study will be conducted are S (Strong) and W (weak) conditions. Management's best estimates for net profits (in millions of dollars) is given below. What strategy should the company do?

Decision	Strong (with Probability 45%)	Weak (with Probability 55%)
A	30	−8
B	20	7
C	5	15

2.8 We have a choice of two investment strategies: stocks and bonds. The returns for each under two possible economic conditions are as follows:

States of Nature

$$p1 = 0.75 \quad p2 = 0.25$$

Alternative	Condition 1	Condition 2
Stocks	$10,000	−$4,000
Bonds	$7,000	$2,000

a. Compute the expected value and select the best alternative.

b. What probabilities for condition 1 and condition 2 would have to exist to be indifferent toward stocks and bonds?

2.9 Company ABC has developed a new line of products. Top management is attempting to decide on both marketing and production strategies. Three strategies are considered, and will be referred to as A(aggressive), B (basic), and C (cautious). The conditions under which the study will be conducted are S (Strong) and W (weak) conditions. Management's best estimates for net profits (in millions of dollars) is given below. Build a decision tree to assist the company to determine the best strategy.

Decision	Strong (with Probability 45%)	Weak (with Probability 55%)
A	30	−8
B	20	7
C	5	15

2.10 An oil company is considering making a bid on a shale oil development contract to be awarded by the government. The company has decided to bid $210 million. They estimate that they have a 70% chance of winning the contract bid. If the company wins the contract, the management has three alternatives for processing the shale. It can develop a new method for processing the oil, use the existing method, or ship the shale overseas for processing. The development cost of the new process is estimated at $30 million. The outcomes and probabilities associated with developing the new method is given below:

Event	Probability	Financial Outcome (in Millions)
Extremely successful	0.7	$450
Moderately successful	0.2	$200
Failure	0.1	$9

The existing methods cost $7 million to execute and the outcomes and probabilities are given below:

Event	Probability	Financial Outcome (in Millions)
Extremely successful	0.6	$300
Moderately successful	0.2	$200
Failure	0.2	$40

The cost to ship overseas is $5 million. If it is shipped overseas, the contract guarantee is $230 million. Construct a decision tree and determine the best strategy.

2.11 The US Army recruiting office has $150,000 and wants to decide whether to market a new recruitment strategy. They have three alternatives.

Alt. 1: Test locally then utilize the results of the local study to determine if a national study is needed.

Alt. 2: Immediately market with no studies.

Alt. 3: Immediately decide not to use the new strategy and keep everything "status quo".

In the absence of a study, they believe the new strategy has a 55% chance of success and a 45% chance of failure at the national level. If successful, the new strategy will bring $300,000 additional assets and if failure, we will lose $100,000 in assets. If they do the study (which costs $30,000), there is a 60% chance of favorable outcome and a 40% chance of an unfavorable outcome. If the study shows that it is a local success, then there is an 85% chance that it will be a national success. If the study shows it was a local failure, there is only a 10% chance that it will be a national success.

What should the recruitment office do?

2.12 Consider the wheel in example 1 with the following probabilities:

Payoff	Probability
$0	.25
$4	.35
$6	.28
$8	.12

Again consider three spins and decide your optimal strategy.

2.13 An art dealer has a client who will buy a masterpiece for $50,000. The dealer can buy the painting today for $40,000 (making a $10,000 profit). Alternatively, the dealer can wait one day when the price is expected to go down to $30,000 or wait another day until the cost goes down to $25,000. After that the painting will no longer available. On each day, there is a 2/3 chance that the paining will be sold to someone else and no longer available to the dealer.

a. Draw a decision tree representing the dealer's decision-making process.

b. Solve the tree. What is dealer's expected profit? When should be buy the painting?

2.14 We are considering one of three alternatives A, B, or C under uncertain conditions. The payoff matrix is below:

	Conditions		
Alternative	#1	#2	#3
A	3,000	4,500	6,000
B	1,000	9,000	2,000
C	4,500	4,000	3,500

Determine the best plan by each of the following criteria and show work:

a. Laplace

b. Maximin

c. Maximax

d. Hurwicz (assume that $\alpha = .65$)

e. Regret (minimax)

f. We have a choice of two investment strategies: stocks and bonds. The returns for each under two possible economic conditions are as follows:

States of Nature

$$p1 = 0.75 \quad p2 = 0.25$$

Alternative	Condition 1	Condition 2
Stocks	$10,000	-$4,000
Bonds	$7,000	$2,000

a. Compute the expected value and select the best alternative.

b. What probabilities for condition 1 and condition 2 would have to exist to be indifferent toward stocks and bonds.

2.15 Given the following payoff matrix:

	Conditions		
Alternative	#1	#2	#3
A	$1,000	2,000	500
B	800	1,200	900
C	700	700	700

Determine the best plan by each of the following criteria and show work:

a. Laplace

b. Maximin

c. Maximax

d. Hurwicz (assume $\alpha = 0.55$)

e. Regret

Reference

Fox, W. (2018). *Mathematical Modeling for Business Analytics*. Boca Raton, FL: Taylor and Francis.

3

Decisions under Certainty: Mathematical Programming Modeling: Linear, Integer, and Mixed Integer Optimization

3.1 Introduction

Consider the Emergency Service Coordinator (ESC) for a county that is interested in locating the county's three ambulances to maximize the residents that can be reached within 8 minutes in emergency situations. The county is divided into six zones and the average time required to travel from one region to the next under semi-perfect conditions are summarized in Table 3.1 and populations in each zone in Table 3.2.

The population in zones 1 to 6 is given in Table 3.2.

Problem Identification: Determine the location for placement of the ambulances to maximize coverage within the allotted time.

Assumptions: Time travel between zones is negligible. Times in the data are averages under ideal circumstances.

Here we further assume that employing an optimization technique would be worthwhile. We will start with assuming a linear model and then we will enhance the model with integer programming.

TABLE 3.1

Average Travel Times from Zone i to Zone j in Perfect Conditions

	1	2	3	4	5	6
1	1	8	12	14	10	16
2	8	1	6	18	16	16
3	12	18	1.5	12	6	4
4	16	14	4	1	16	12
5	18	16	10	4	2	2
6	16	18	4	12	2	2

DOI: 10.1201/9781032726885-3

TABLE 3.2

Populations in Each Zone

1	50,000
2	80,000
3	30,000
4	55,000
5	35,000
6	20,000
Total	270,000

Perhaps, consider planning the shipment of needed items from the warehouses where they are manufactured and stored to the distribution centers where they are needed.

There are three warehouses at different cities: Detroit, Pittsburgh, and Buffalo. They have 250, 130, and 235 tons of paper accordingly. There are four publishers in Boston, New York, Chicago, and Indianapolis. They ordered 75, 230, 240, and 70 tons of paper to publish new books. Table 3.3 provides the costs in dollars of transportation of one ton of paper.

Management wants you to minimize the shipping costs while meeting demand. This problem involves the allocation of resources and can be modeled as a linear programming (LP) problem as we will discuss.

In engineering management, the ability to optimize results in a constrained environment is crucial to success. Additionally, the ability to perform critical sensitivity analysis or "what if analysis" is extremely important for decision-making. Consider starting a new diet to be healthy. You go to a nutritionist that gives you lots of information on foods. They recommend sticking to six different foods – bread, milk, cheese, fish, potato, and yogurt – and provides you a table of information including the average cost of the item (Table 3.4).

We go to a nutritionist and she recommends that our diet contains not less than **150** calories, not more than **10** g of protein, not less than **10** g of carbohydrates, and not less than **8** g of fat. Also, we decide that our diet should have **minimal cost**. In addition, we conclude that our diet should include at least **0.5** g of fish and not more than **1** cups of milk. Again this is an allocation of recourses problem where we want the optimal diet at minimum cost. We have six unknown variables that define weight of the food. There is a lower

TABLE 3.3

Costs for Shipping

From\To	Boston (BS)	New York (NY)	Chicago (CH)	Indianapolis (IN)
Detroit (DT)	15	20	16	21
Pittsburgh (PT)	25	13	5	11
Buffalo (BF)	15	15	7	17

TABLE 3.4

Information Table

	Bread	Milk	Cheese	Potato	Fish	Yogurt
Cost ($)	2.0	3.5	8.0	1.5	11.0	1.0
Protein (g)	4.0	8.0	7.0	1.3	8.0	9.2
Fat (g)	1.0	5.0	9.0	0.1	7.0	1.0
Carbohydrates (g)	15.0	11.7	0.4	22.6	0.0	17.0
Calories (Cal)	90	120	106	97	130	180

bound for fish as 0.5 g. There is an upper bound for milk as 1 cup. To model and solve this problem, we can use LP.

Modern LP was the result of a research project undertaken by the US Department of Air Force under the title of Project SCOOP (Scientific Computation of Optimum Programs). As the number of fronts in the Second World War increased, it became more and more difficult to coordinate troop supplies effectively. Mathematicians looked for ways to use the new computers being developed to perform calculations quickly. One of the SCOOP team members, George Dantzig, developed the simplex algorithm for solving simultaneous LP problems. The simplex method has several advantageous properties: it is very efficient, allowing its use for solving problems with many variables; it uses methods from linear algebra, which are readily solvable.

In January 1952, the first successful solution to an LP problem was found using a high-speed electronic computer on the National Bureau of Standards SEAC machine. Today, most LPs are solved via high-speed computers. Computer-specific software, such as LINDO, EXCEL SOLVER, GAMS, have been developed to help in the solving and analysis of LP problems. We will use the power of LINDO to solve our LP problems in this chapter.

To provide a framework for our discussions, we offer the following basic model:

Maximize (or minimize) $f(X)$

Subject to:

$$g_i(X) \begin{cases} \geq \\ = \\ \leq \end{cases} b_i \text{ for all } i.$$

Now let's explain this notation. The various component of the vector X are called the decision variables of the model. These are the variables that can be controlled or manipulated. The function, $f(X)$, is called the objective function. By subject to, we connote that there are certain side conditions, resource requirement, or resource limitations that must be met. These conditions are called constraints. The constant b_i represents the level that the associated constraint $g(X_i)$ and is called the right-hand side in the model.

LP is a method for solving linear problems, which occur very frequently in almost every modern industry. In fact, areas using LP are as diverse as defense, health, transportation, manufacturing, advertising, and telecommunications. The reason for this is that in most situations, the classic economic problem exists – you want to maximize output but are competing for limited resources. The "Linear" in Linear Programming means that in the case of production, the quantity produced is proportional to the resources used and also the revenue generated. The coefficients are constants and no products of variables are allowed.

In order to use this technique, the company must identify a number of constraints that will limit the production or transportation of their goods; these may include factors such as labor hours, energy, and raw materials. Each constraint must be quantified in terms of one unit of output, as the problem solving method relies on the constraints being used.

An optimization problem that satisfies the following five properties is said to be an LP problem.

- There is a unique objective function, $f(X)$.
- Whenever a decision variable, X, appears in either the objective function or a constraint function, it must appear with an exponent of 1, possibly multiplied by a constant.
- No terms contain products of decision variables.
- All coefficients of decision variables are constants and know with certainty.
- Decision variables are permitted to assume fractional as well as integer values.

Linear problems, by the nature of the many unknowns, are very hard to solve by human inspection, but methods have been developed to use the power of computers to do the hard work.

3.2 Formulating Linear Programming Problems

An LP problem is a problem that requires an objective function to be maximized or minimized subject to resource constraints. The key to formulating an LP problem is recognizing the decision variables. The objective function and all constraints are written in terms of these decision variables.

The conditions for a mathematical model to be a linear program (LP) were:

- all variables continuous (i.e., can take fractional values)
- a single objective (minimize or maximize)
- the objective and constraints are linear, i.e., any term is either a constant or a constant known with certainty multiplied by an unknown
- the decision variables must be non-negative.

LPs are important – this is because:

- many practical problems can be formulated as LPs
- there exists an algorithm (called the *simplex* algorithm) that enables us to solve LPs numerically relatively easily.

We will return later to the simplex algorithm for solving LPs but for the moment we will concentrate upon formulating LPs.

Some of the major application areas to which LP can be applied are:

- Blending
- Production planning
- Oil refinery management
- Distribution
- Financial and economic planning
- Manpower planning
- Blast furnace burdening
- Farm planning

We consider below some specific examples of the types of problem that can be formulated as LPs. Note here that the key to formulating LPs is **practice**. However, a useful hint is that common objectives for LPs are **minimize cost/maximize profit**.

Example 3.1: Manufacturing

Consider the following problem statement. A company wants to can two new different drinks for the holiday season. It takes 2 hours to can one gross of Drink A, and it takes 1 hour to label the cans. It takes 3 hours to can one gross of Drink B, and it takes 4 hours to label the cans. The company makes $10 profit on one gross of Drink A and a $20 profit of one gross of Drink B. Given that we have 20 hours to devote to canning the drinks and 15 hours to devote to labeling cans per week, how many cans of each type drink should the company package to maximize profits?

Problem Identification: Maximize the profit of selling these new drinks. Define variables:

X_1 = the number of gross cans produced for Drink A per week

X_2 = the number of gross cans produced for Drink B per week

Objective Function:

$$Z = 10X_1 + 20X_2$$

Constraints:

1. Canning with only 20 hours available per week

$$2X_1 + 3X_2 \leq 20$$

2. Labeling with only 15 hours available per week

$$X_1 + 4X_2 \leq 15$$

3. Non-negativity restrictions

$$X_1 \geq 0 \left(\text{non-negativity of the production items}\right)$$

$$X_2 \geq 0 \left(\text{non-negativity of the production items}\right)$$

The Complete FORMULATION:

Maximize $Z = 10X_1 + 20X_2$

Subject to:
$2X_1 + 3X_2 \leq 20$
$X_1 + 4X_2 \leq 15$
$X_1 \geq 0$
$X_2 \geq 0$

We will see in the next section how to solve these two-variable problems graphically.

Example 3.2: Financial Planning

A bank makes four kinds of loans to its personal customers and these loans yield the following annual interest rates to the bank:

- First mortgage 14%
- Second mortgage 20%
- Home improvement 20%
- Personal overdraft 10%

The bank has a maximum foreseeable lending capability of $250 million and is further constrained by the policies:

1. first mortgages must be at least 55% of all mortgages issued and at least 25% of all loans issued (in $ terms)
2. second mortgages cannot exceed 25% of all loans issued (in $ terms)
3. to avoid public displeasure and the introduction of a new windfall tax, the average interest rate on all loans must not exceed 15%.

Formulate the bank's loan problem as an LP so as to maximize interest income while satisfying the policy limitations.

Note here that these policy conditions, while potentially limiting the profit that the bank can make, also limit its exposure to risk in a particular area. It is a fundamental principle of risk reduction that risk is reduced by spreading money (appropriately) across different areas.

3.2.1 Financial Planning Formulation

Note here that, as in *all* formulation exercises, we are translating a verbal description of the problem into an *equivalent* mathematical description.

A useful tip when formulating LPs is to express the variables, constraints, and objective in words before attempting to express them in mathematics.

3.2.1.1 Variables

Essentially we are interested in the amount (in dollars) the bank has loaned to customers in each of the four different areas (not in the actual number of such loans). Hence, let

x_i = amount loaned in area i in millions of dollars (where $i = 1$ corresponds to first mortgages, $i = 2$ to second mortgages (etc.) and note that each $x_i \geq = 0$ ($i = 1,2,3,4$).

Note here that it is conventional in LPs to have all variables $> = 0$. Any variable (X, say) which can be positive *or* negative can be written as $X_1 - X_2$ (the difference of two new variables) where $X_1 > = 0$ and $X_2 > = 0$.

3.2.1.2 Constraints

a. limit on amount lent

$$x_1 + x_2 + x_3 + x_4 \leq 250$$

b. policy condition 1

$$x_1 \geq 0.55(x_1 + x_2)$$

i.e., first mortgages > = 0.55 (total mortgage lending) and also

$$x_1 \geq 0.25(x_1 + x_2 + x_3 + x_4)$$

i.e., first mortgages ≥ 0.25 (total loans)
c. policy condition 2

$$x_2 \leq 0.25(x_1 + x_2 + x_3 + x_4)$$

d. policy condition 3 – we know that the total annual interest is $0.14x_1 + 0.20x_2 + 0.20x_3 + 0.10x_4$ on total loans of $(x_1 + x_2 + x_3 + x_4)$. Hence, the constraint relating to policy condition (3) is

$$0.14x_1 + 0.20x_2 + 0.20x_3 + 0.10x_4 \leq 0.15(x_1 + x_2 + x_3 + x_4)$$

3.2.1.3 Objective Function

To maximize interest income (which is given above), i.e.,

$$\text{Maximize } Z = 0.14x_1 + 0.20x_2 + 0.20x_3 + 0.10x_4$$

Example 3.3: Blending and Formulation

Consider the example of a manufacturer of animal feed who is producing feed mix for dairy cattle. In our simple example, the feed mix contains two active ingredients. One kilogram of feed mix must contain a minimum quantity of each of four nutrients as below:

Nutrient	A	B	C	D
Gram	90	50	20	2

The ingredients have the following nutrient values and cost

	A	B	C	D	Cost/kg
Ingredient 1 (gram/kg)	100	80	40	10	40
Ingredient 2 (gram/kg)	200	150	20	0	60

What should be the amounts of active ingredients in 1 kg of feed mix that minimizes cost?

3.2.2 Blending Problem Solution

3.2.2.1 Variables

In order to solve this problem, it is best to think in terms of 1 kg of feed mix. That kilogram is made up of two parts – ingredient 1 and ingredient 2:

x_1 = amount (kg) of ingredient 1 in 1 kg of feed mix

x_2 = amount (kg) of ingredient 2 in 1 kg of feed mix

where $x_1 \geq 0, x_2 \geq 0$

Essentially, these variables (x_1 and x_2) can be thought of as the recipe telling us how to make up 1 kg of feed mix.

3.2.2.2 Constraints

- nutrient constraints

$$100x_1 + 200x_2 >= 90 \ (nutrient \ A)$$

$$80x_1 + 150x_2 >= 50 \ (nutrient \ B)$$

$$40x_1 + 20x_2 >= 20 \ (nutrient \ C)$$

$$10x_1 >= 2 \ (nutrient \ D)$$

- balancing constraint (an *implicit* constraint due to the definition of the variables)

$$x_1 + x_2 = 1$$

3.2.2.3 Objective Function

Presumably to minimize cost, i.e.,

$$\text{Minimize } Z = 40x_1 + 60x_2$$

This gives us our complete LP model for the blending problem.

Example 3.4: Production planning problem

A company manufactures four variants of the same table and in the final part of the manufacturing process, there are assembly, polishing, and packing operations. For each variant, the time required for these operations is shown below (in minutes) as is the profit per unit sold.

Variant	Assembly	Polish	Pack	Profit ($)
1	2	3	2	1.50
2	4	2	3	2.50
3	3	3	2	3.00
4	7	4	5	4.50

- Given the current state of the labor force, the company estimate that, each year, they have 100,000 minutes of assembly
  time, 50,000 minutes of polishing time, and 60,000 minutes
  of packing time available. How many of each variant
  should the company make per year and what is the associated profit?

3.2.2.4 Variables

Let:

x_i be the number of units of variant i ($i =1,2,3,4$) made per year

where $x_i \geq 0$, $i =1,2,3,4$

3.2.2.5 Constraints

Resources for the operations of assembly, polishing, and packing

$$2x_1 + 4x_2 + 3x_3 + 7x_4 <= 100,000 \ (assembly)$$

$$3x_1 + 2x_2 + 3x_3 + 4x_4 <= 50,000 \ (polishing)$$

$$2x_1 + 3x_2 + 2x_3 + 5x_4 <= 60,000 \ (packing)$$

3.2.2.6 Objective function

$$\text{Maximize } Z = 1.5x_1 + 2.5x_2 + 3.0x_3 + 4.5x_4$$

Example 3.5: Shipping

Consider planning the shipment of needed items from the warehouses where they are manufactured and stored to the distribution
centers where they are needed as shown in the introduction. There
are three warehouses at different cities: Detroit, Pittsburgh, and
Buffalo. They have 250, 130, and 235 tons of paper, respectively. There
are four publishers in Boston, New York, Chicago, and Indianapolis.
They ordered 75, 230, 240, and 70 tons of paper, respectively, to publish new books.

There are the following costs in dollars of transportation of one ton of paper:

From\To	Boston (BS)	New York (NY)	Chicago (CH)	Indianapolis (IN)
Detroit (DT)	15	20	16	21
Pittsburgh (PT)	25	13	5	11
Buffalo (BF)	15	15	7	17

Management wants you to minimize the shipping costs while meeting demand.

We define x_{ij} to be the travel from city i (1 is Detroit, 2 is Pittsburg, 3 is Buffalo) to city j (1 is Boston, 2 is New York, 3 is Chicago, and 4 is Indianapolis).

$$\text{Minimize } Z = 15x_{11} + 20x_{12} + 16x_{13} + 21x_{14} + 25x_{21} + 13x_{22} + 5x_{23} + 11x_{24} + 15x_{31} + 15x_{32} + 7x_{33} + 17x_{34}$$

Subject to:
$$x_{11} + x_{12} + x_{13} + x_{14} \leq 250 \ (\textit{availability in Detroit})$$
$$x_{21} + x_{22} + x_{23} + x_{24} \leq 130 \ (\textit{availability in Pittsburg})$$
$$x_{31} + x_{32} + x_{33} + x_{34} \leq 235 \ (\textit{availability in Buffalo})$$
$$x_{11} + x_{21} + x_{31} \geq 75 \ (\textit{demand Boston})$$
$$x_{12} + x_{22} + x_{32} \geq 230 \ (\textit{demand New York})$$
$$x_{13} + x_{23} + x_{33} \geq 240 \ (\textit{demand Chicago})$$
$$x_{14} + x_{24} + x_{34} \geq 70 \ (\textit{demand Indianapolis})$$
$$x_{ij} \geq 0$$

3.2.3 Integer Programming

For integer programming, we will take advantage of technology. We will not present the branch and bound technique but suggest a thorough review of the topic can be found in Winston or other similar math programming texts. We will only illustrate the solution via technology. For mixed integer programming, we show the formulations and the use of technology.

3.2.4 Nonlinear Programming

It is not our plan to present material here on how to formulate or solve nonlinear programs in Nash equilibriums, Chapter 8, or Nash Arbitration, Chapter 11. Often, we have nonlinear objective functions or constraints. Suffice it to say, we will recognize these and use technology to assist in the solution. Excellent nonlinear programming can be read for additional information.

3.3 Graphical Linear Programming

Many applications in business and economics involve a process called optimization. In optimization problems, you are asked to find the minimum or the maximum result. This section illustrates the strategy in graphical simplex of LP. We will restrict ourselves in this graphical context to two dimensions. Variables in the simplex method are restricted to positive variables (e.g., $x \geq 0$).

A two-dimensional LP problem consists of a linear objective function and a system of linear inequalities called resource constraints. The objective function gives the linear quantity that is to be maximized (or minimized). The constraints determine the *set of feasible solutions*. We find this section useful when we cover William's method (graphical) in Chapter 5. Let's illustrate graphical LP.

3.3.1 Memory Chips for CPUs

Let's start with a manufacturing example. Suppose a small business wants to know how many of two types of high-speed computer chips to manufacturer weekly to maximize their profits. First, we need to define our decision variables. Let

x_1 = number of high speed chip type A to produce weekly

x_2 = number of high speed chip type B to produce week

The company reports a profit of \$140 for each type *A* chip and \$120 for each type *B* chip sold. The production line reports the following information:

	Chip A	Chip B	Quantity Available
Assembly time (hours)	2	4	1,400
Installation time (hours)	4	3	1,500
Profit (per unit)	140	120	

The constraint information from the table becomes inequalities that are written mathematical as

$$2x_1 + 4x_2 \leq 1,400 \ (\text{assembly time})$$

$$4x_1 + 3x_2 \leq 1,500 \ (\text{installation time})$$

$$x_1 \geq 0, x_2 \geq 0$$

The profit equation is:

$$\text{Profit } Z = 140x_1 + 120x_2$$

3.3.2 The Feasible Region

We use the constraints of the linear program:

$$2x_1 + 4x_2 \le 1,400 \text{ (assembly time)}$$
$$4x_1 + 3x_2 \le 1,500 \text{ (installation time)}$$
$$x_1 \ge 0, x_2 \ge 0$$

The constraints of a linear program, which include any bounds on the decision variables, essentially shape the region in the x–y plane that will be the domain for the objective function prior to any optimization being performed. Every inequality constraint that is part of the formulation divides the entire space defined by the decision variables into two parts: the portion of the space containing points that violate the constraint, and the portion of the space containing points that satisfy the constraint.

It is very easy to determine which portion will contribute to shaping the domain. We can simply substitute the value of some point in either *half-space* into the constraint. Any point will do, but the origin is particularly appealing. Since there's only one origin, if it satisfies the constraint, then the *half-space* containing the origin will contribute to the domain of the objective function.

When we do this for each of the constraints in the problem, the result is an area representing the intersection of all the *half-spaces* that satisfied the constraints individually. This intersection is the domain for the objective function for the optimization. Because it contains points that satisfy all the constraints simultaneously, these points are considered feasible to the problem. The common name for this domain is the *feasible region*.

Consider our constraints:

$$2x_1 + 4x_2 \le 1,400 \text{ (assembly time)}$$
$$4x_1 + 3x_2 \le 1,500 \text{ (installation time)}$$
$$x_1 \ge 0, x_2 \ge 0$$

For our graphical work, we use the constraints: $x_1 \ge 0$, $x_2 \ge 0$ to set the region. Here, we are strictly in the x_1–x_2 plane (the first quadrant).

Let's first take constraint #1 (assembly time) in the first quadrant: $2x_1 + 4x_2 \le 1,400$ shown in Figure 3.1.

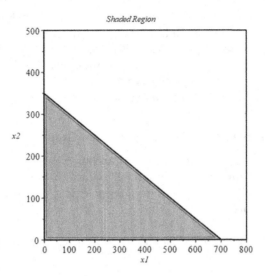

FIGURE 3.1
Shaded inequality.

We graph each constraint as an equality, one at a time. We choose a point, usually the origin to test the validity of the inequality constraint. We shade all the areas where the validity holds. We repeat this process for all constraints to obtain Figure 3.2.

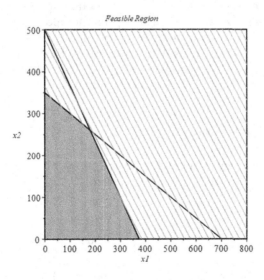

FIGURE 3.2
Plot of (1) the assembly hour's constraint and (2) the installation hour's constraint in the first quadrant.

Figure 3.2 shows a plot of (1) the assembly hour's constraint and (2) the installation hour's constraint in the first quadrant. Along with the non-negativity restrictions on the decision variables, the intersection of the half-spaces defined by these constraints is the feasible region shown in gray. This area represents the domain for the objective function optimization.

We region shaded in our feasible region.

3.3.3 Solving a Linear Programming Problem Graphically

We have decision variables defined and an objection function that is to be maximized or minimized. Although all points inside the feasible region provide feasible solutions, the solution, if one exists, occurs according to the Fundamental Theorem of LP:

> *If the optimal solution exists, then it occurs at a corner point of the feasible region*

Note the various corners formed by the intersections of the constraints in example. These points are of great importance to us. There is a cool theorem (didn't know there were any of these, huh?) in linear optimization that states, "if an optimal solution exists, then an optimal corner point exists". The result of this is that any algorithm searching for the optimal solution to a linear program should have some mechanism of heading toward the corner point where the solution will occur. If the search procedure stays on the outside border of the feasible region while pursuing the optimal solution, it is called an *exterior point* method. If the search procedure cuts through the inside of the feasible region, it is called an *interior point* method.

Thus, in an LP problem, if there exists a solution, it must occur at a corner point of the set of feasible solutions (these are the vertices of the region). Note that in Figure 3.2 the corner points of the feasible region are the four coordinates and we might use algebra to find these: (0,0), (0,350) (375, 0), and (180,260).

How did we get the point (180,260)?

This point is the intersection of the lines: $2x + 4y = 1,400$ and $4x+3y = 1,500$. We use matrix algebra and solve for (x,y) from

$$\begin{bmatrix} 2 & 4 \\ 4 & 3 \end{bmatrix} \begin{bmatrix} x \\ y \end{bmatrix} = \begin{bmatrix} 1,400 \\ 1,500 \end{bmatrix}$$

Now, that we have all the possible solution coordinates for (x, y), we need to know which is the optimal solution. We evaluate the objective function at each point and choose the best solution.

Assume our objective function is to Maximize $Z = 2x + 2y$. We can set up a table of coordinates and corresponding Z-values as follows.

Coordinate of Corner Point	$Z = 140x + 120y$
(0,0)	$Z = 0$
(0,350)	$Z = 42,000$
(180,260)	$Z = 56,400$(Best answer)
(375,0)	$Z = 52,500$
Best solution is (180,260)	$Z = 56,400$

Graphically, we see the result by plotting the objective function line, $Z = 2x + 2y$, with the feasible region. Determine the parallel direction for the line to maximize (in this case) Z. Move the line parallel until it crosses the last point in the feasible set. That point is the solution. The line that goes through the origin at a slope of $-2/2$ is called the ISO-Profit line. We have provided this in Figure 3.3.

We summarize the steps for solving an LP problem involving only two variables.

1. Sketch the region corresponding to the system of constraints. The points satisfying all constraints make up the feasible solution.
2. Find all the corner points (or intersection points in the feasible region).

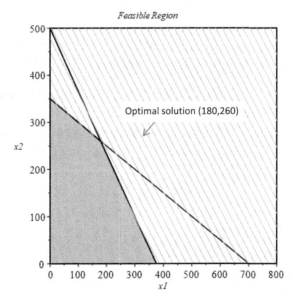

FIGURE 3.3
Iso-profit lines added.

3. Test the objective function at each corner point and select the values of the variables that optimize the objective function. For bounded regions, both a maximum and a minimum will exist. For an unbounded region, if a solution exists, it will exist at a corner.

3.4 Linear Programming with Technology

Technology is critical to solving, analyzing, and performing sensitivity analysis on LP problems. Technology provides a suite of powerful, robust routines for solving optimization problems, including LPs. Technology that we illustrate include Excel, LINDO, and LINGO as these appear to be used often in engineering. We also examined GAMS, which we found powerful but too cumbersome to discuss here. We tested all these other software packages and found them useful.

First we present a 'How to" in Excel.

Consider the following example:

MAXIMIZE $2X1 + 6X2 + 5X3$

Subject to:

$$X1 + X2 + X3 = 3$$
$$X1 + 2X2 + 3X3 \leq 10$$
$$2X1 + 6X2 + X3 \geq 5$$
$$X1, X2, X3 \geq 0$$

Using two variable models, we can graph it on paper and easily solve for the solution. With <u>more</u> than two variables (X1, X2, X3), LP must be used and this example will demonstrate how to use MS Excel in that capacity.

- First, open MS Excel, and type in/set-up an LP that looks exactly like the one below:

Key: RHS = Right-Hand Side (for constraints)
Const 1 = Constraint #1 (same for Const 2 and 3)
Dec vars = Decision Variables (the values for X1, X2, and X3 we seek)

To this point, the Linear Program is all "just typing", with no blocks identified with formulas in them. A title is input in cell A1 – call it whatever you want, in this case "Linear Program Example" – and the spacings between cells are merely to make it an "easy to read" format up to this point. In future problems, you can label the constraints and Decision Variables any way you would like – Instead of "Const 1", you can use "Labor Hours", or "Amount of Material Available", etc. instead of Dec Vars, you can use "Number of fixtures to produce", etc. You can use any terminology you like, as long as YOU understand it – just remember that Dec Vars in Row 10 are the values of X1, X2, and X3 that you seek.

Next, add in the numbers for your problem in the correct columns and rows. In Row G, add in the = and/or < = constraints, then the RHS values in Column H.

At this point, there are two things you can do to make your program look better and make it easier to read and understand: first, highlight the block of cells B3 to B10 to H3 to H10 and Click on "Center", for aesthetics and ease of data presentation; second, color in the Objective Function row, the Constraints rows, and the Dec Vars with different colors to offset them. Starting to look more like a model now!

Now, we need to define cells. In this example, cells F4, F6, F7, and F8 need to be defined. Also, in cells B10, C10, and D10, place the number "1" in each for reasons that will become clear later (briefly, they are to test that the answer cells F4, F6, F7, and F8 have been defined properly). In cells F4, F6, F7, and F8, the "=" sign before the equation tells the cell that this is a mathematical formula to be computed.

Cell F4: This cell will contain the final answer to the Value of the Objective Function. Highlight the cell F4, and type in the following: = B4*B10 + C4*C10 + D4*D10, which signifies 2*X1 + 6*X2 + 5*X3. After completion, the answer in F4 should be "13" at this point, since the Dec Vars are set to 1,1,1, ensuring your coding is correct. If your answer is not 13, re-check your coding in F4.

Cell F6: This cell contains the answer to the first constraint. Highlight cell F6, and type in the following:

= B6*B10 + C6 * C10 + D6* D10, which signifies 1*X1 + 1*X2 + 1*X3. Click on Enter – the answer in F6 should be "3". Don't forget the "=" sign!

Cell F7: This cell contains the answer to the second constraint. Highlight cell F7, and type in the following:

= B7*B10 + C7 * C10 + D7* D10, which signifies 1*X1 + 2*X2 + 3*X3. Click on Enter: the answer in F7 should be "6".

Cell F8: This cell contains the answer to the third constraint. Highlight cell F8, and type in the following:

= B8*B10 + C8 * C10 + D8* D10, which signifies 2*X1 + 6*X2 + 1*X3. Click on Enter: the answer in F8 should be "9".

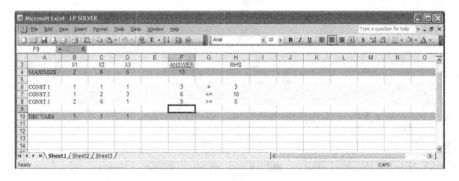

Now we are ready to use the Solver. Save what you have, then go to "Tools" and then select "Solver". (If you can't find Solver, it will be in Add-ins – extract it from there).

1. Set Target Cell: Input F4
2. Select "Max" button
3. In the "By Changing Cells" area, merely highlight all three cells in B10, C10, and D10 with the mouse.
4. Now move to the constraints. Click on "Add".
   In the first constraint: Type in F6, select =, and type in the number 3, in the three areas. Click on OK, then Add ….
   In the second constraint: Type in F7, select < =, type in 10. Click on OK, then Add ….
   In the third constraint: Type in F8, select > =, type in 5. Click on OK

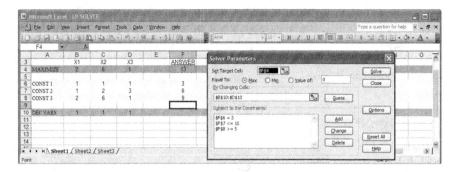

5. Now go to Options: Select "Assume Linear Model", "Assume Non-Negative", "Use Auto Scaling"

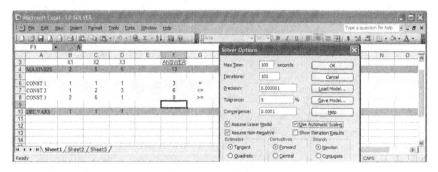

6. Click on OK, then click on Solve.

Answer should now be revealed:

Objective Function = **18** in Cell F4

Constraints are satisfied: **3 = 3**

$$6 < = 10$$
$$18 > = 5$$

Decision Variables: **X1 = 0, X2 = 3, X3 = 0**

Thus, the problem is maximized by making three X2's only, with an objective value profit of 18. Now we present our previous example solved via each technology.

Maximize $Z = 25x_1 + 30x_2$

Subject to:

$20x_1 + 30x_2 \leq 690$

$5x_1 + 4x_2 \leq 120$

$x1, x2, \geq 0$

Using EXCEL

	A	B	C	D	E	F
1	LP in EXCEL					
2						
3						
4	Decision	Variables			Objective Function	
5		Initial/Final Values			=28*B6+30*B7	
6	x1	0				
7	x2	0				
8						
9						

ont   Alignment

=2*B6+3*B7

	A	B	C	D	E	F
1	LP in EXCEL					
2						
3						
4	Decision	Variables			Objective Function	
5		Initial/Final Values			0	
6	x1	0				
7	x2	0				
8						
9						
10	Constraints			Used	RHS	
11				0	100	
12				0	120	
13				0	90	
14						
15						

Solver

Constraints into solver

We now have the Full Set UP. We can click on <u>S</u>olve.

	A	B	C	D	E	F
1	LP in EXCEL					
2						
3						
4	Decision	Variables			Objective Function	
5		Initial/Final Values			972	
6	x1	9				
7	x2	24				
8						
9						
10	Constraints			Used	RHS	
11				90	100	
12				120	120	
13				90	90	
14						

Obtain the answers as $x1 = 9$, $x2 = 24$, $Z = 972$.

Additionally, we can obtain reports from Excel. Two key reports are the answer report and the sensitivity report.

Answer Report

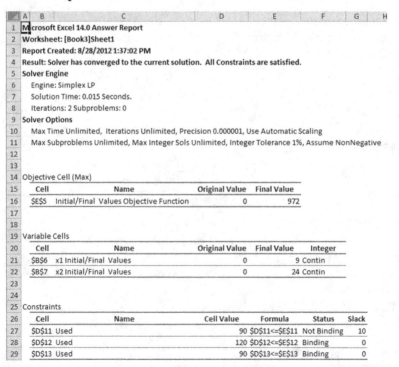

	A	B	C	D	E	F	G	H
1		Microsoft Excel 14.0 Answer Report						
2		Worksheet: [Book3]Sheet1						
3		Report Created: 8/28/2012 1:37:02 PM						
4		Result: Solver has converged to the current solution. All Constraints are satisfied.						
5		Solver Engine						
6		Engine: Simplex LP						
7		Solution Time: 0.015 Seconds.						
8		Iterations: 2 Subproblems: 0						
9		Solver Options						
10		Max Time Unlimited, Iterations Unlimited, Precision 0.000001, Use Automatic Scaling						
11		Max Subproblems Unlimited, Max Integer Sols Unlimited, Integer Tolerance 1%, Assume NonNegative						
12								
13								
14		Objective Cell (Max)						
15		Cell	Name	Original Value	Final Value			
16		$E$5	Initial/Final Values Objective Function	0	972			
17								
18								
19		Variable Cells						
20		Cell	Name	Original Value	Final Value	Integer		
21		$B$6	x1 Initial/Final Values	0	9	Contin		
22		$B$7	x2 Initial/Final Values	0	24	Contin		
23								
24								
25		Constraints						
26		Cell	Name	Cell Value	Formula	Status	Slack	
27		$D$11	Used	90	$D$11<=$E$11	Not Binding	10	
28		$D$12	Used	120	$D$12<=$E$12	Binding	0	
29		$D$13	Used	90	$D$13<=$E$13	Binding	0	

Sensitivity Report

	A	B	C	D	E	F	G	H
1		Microsoft Excel 14.0 Sensitivity Report						
2		Worksheet: [Book3]Sheet1						
3		Report Created: 8/28/2012 1:37:03 PM						
4								
5								
6		Variable Cells						
7				Final	Reduced	Objective	Allowable	Allowable
8		Cell	Name	Value	Cost	Coefficient	Increase	Decrease
9		$B$6	x1 Initial/Final Values	9	0	28	6.285714286	8
10		$B$7	x2 Initial/Final Values	24	0	30	12	5.5
11								
12		Constraints						
13				Final	Shadow	Constraint	Allowable	Allowable
14		Cell	Name	Value	Price	R.H. Side	Increase	Decrease
15		$D$11	Used	90	0	100	1E+30	10
16		$D$12	Used	120	4.8	120	60	15
17		$D$13	Used	90	4.4	90	10	30
18								

We find our solution is $x_1 = 9$, $x_2 = 24$, $P = \$972$. From the standpoint of sensitivity analysis, Excel is satisfactory in that it provides shadow prices.

Limitation: No tableaus are provided making it difficult to find alternate solutions.

However, we can use the revised simplex equations as matrices to create the tableau if desired. From Winston (2005), we define the following terms as matrices.

BV = matrix of basic variables

B = right-hand side column matrix

a_j = column matrix for x_j in the jth constraints

$B = m \times m$ matrix

C_j = row matrix of costs.

C_{BV} = original costs of basic variables

Summarizing formulas that we can use,

$$B^{-1}a_j$$

$$C_{BV}B^{-1}a_j - C_j$$

$$B^{-1}b$$

Maximize $Z = 25x_1 + 30x_2$

Subject to:
$$20x_1 + 30x_2 \leq 690$$
$$5x_1 + 4x_2 \leq 120$$
$$x1, x2, \geq 0$$

Set up the LP

z	x1	x2	s1	s2	rhs
1	−25	−30	0	0	0
0	20	30	1	0	690
0	5	4	0	1	120

Step 1. Place this matrix in Excel as shown below in Standard form.

B					
z	x1	x2	s1	s2	rhs
1	−25	−30	0	0	0
0	20	30	1	0	690
0	5	4	0	1	120

Step 2. Determine the entering variables as the cost in row 1 that is most negative. In this case, it is $x2$ at −30.

Step 3. Determines who leaves by performing the minimum positive ratio test. In this case, we compare 690/30 and 120/4. We find 690/30=23 is smaller, so S1 departs.

Step 4. Write the new BV matrix replacing column 2 0,1,0 with the original column of $x2$ −30,30,4.

New Bbv	1	−30	0
	0	30	0
	0	4	1

Step 5. Find the matrix inverse of the new Bbv matrix.

1	1	0
0	0.033333333	0
0	-0.133333333	1

Step 6. Multiple the new Bbv and the original B matrix to obtain the updated tableau.

z	x1	x2	s1	s2	rhs
1	-5	0	1	0	690
0	0.666666667	1	0.033333	0	23
0	2.333333333	0	-0.13333	1	28

Step 7. Check for optimality conditions, all $Cj > = 0$. If not return to Step 2 and repeat steps 2–7 again.

We are not optimal as the coefficient of x1 in the Cj row is-5. We repeat steps 2–7.

	z	x1	x2	s1	s2	rhs
25	z	x1	x2	s1	s2	rhs
26	1	-5	0	1	0	690
27	0	0.666666667	1	0.033333	0	23
28	0	2.333333333	0	-0.13333	1	28
29						
30	x1 enters as -5<0					
31	ratio test	34.5				
32		12				
33	s2 leaves because 12<34.5					
34						
35		New BV		1	-30	-25
36				0	30	20
37				0	4	5
38						
39		B^-1		1	0.714286	2.142857
40				0	0.071429	-0.28571
41				0	-0.05714	0.428571
42						
43		1	-3.55271E-15	-3.55271E-15	0.714286	2.142857	750
44		0	0	1	0.071429	-0.28571	15
45		0	1	0	-0.05714	0.428571	12

We find we are now optimal as all $Cj > = 0$. Our solution is $Z = 750$ when $x_1 = 12, x_2 = 15$.

3.5 Case Studies in Linear Programming

Example 3.6: Supply Chain Operations (from Fox et al., 2014)

In our case study, we present LP for supply chain design decision-making. We consider producing a new mixture of gasoline. We desire to minimize the total cost of manufacturing and distributing the new mixture. There is a supply chain involved with a product that must be modeled. The product is made up of components that are produced separately.

Crude Oil Type	Compound A (%)	Compound B (%)	Compound C (%)	Cost/ Barrel	Barrel Avail (000 of Barrels)
X10	35	25	35	$26	15,000
X20	50	30	15	$32	32,000
X30	60	20	15	$55	24,000

Demand information is as follows:

Gasoline	Compound A (%)	Compound B (%)	Compound C (%)	Expected Demand (000 of Barrels)
Premium	≥ 55	≤ 23		14,000
Super		≥ 25	≤ 35	22,000
Regular	≥ 40		≤ 25	25,000

Let i = crude type 1, 2, 3 (X10,X20,X30, respectively)
Let j = gasoline type 1,2,3 (Premium, Super, Regular, respectively)

We define the following decision variables:

Gij = amount of crude i used to produce gasoline j

For example,

G_{11} = amount of crude X10 used to produce Premium gasoline
G_{12} = amount of crude type X20 used to produce Premium gasoline
G_{13} = amount of crude type X30 used to produce Premium gasoline
G_{12} = amount of crude type X10 used to produce Super gasoline
G_{22} = amount of crude type X20 used to produce Super gasoline
G_{32} = amount of crude type X30 used to produce Super gasoline
G_{13} = amount of crude type X10 used to produce Regular gasoline
G_{23} = amount of crude type X20 used to produce Regular gasoline
G_{33} = amount of crude type X30 used to produce Regular gasoline

LP formulation

Minimize Cost = $86 $(G11 + G21 + G31)$ + 92(G12 + G22 + G32)$
$\qquad$ + 95(G13 + G23 + G33)$

Subject to:
Demand

$G11 + G21 + G31 > 14,000$ (Premium)

$G12 + G22 + G32 > 22,000$ (Super)

$G13 + G23 + G33 > 25,000$ (Regular)

Availability of products

$G11 + G12 + G13 < 15,000$ (crude 1)

$G21 + G22 + G23 < 32,000$ (crude 2)

$G31 + G32 + G33 < 24,000$ (crude 3)

Product mix in mixture format

$(0.35\ G11 + 0.50\ G21 + 0.60\ G31)/(G11 + G21 + G31) \geq 0.55$ (X10 in Premium)

$(0.25\ G11 + 0.30\ G21 + 0.20\ G31)/(G11 + G21 + G31) \leq 0.23$ (X20 in Premium)

$(0.35\ G13 + 0.15\ G23 + 0.15\ G33)/(G13 + G23 + G33) \geq 0.25$ (X20 in Regular)

$(0.35\ G13 + 0.15\ G23 + 0.15\ G33)/(G13 + G23 + G33) \leq 0.35$ (X30 in Regular)

$(0.35G12 + 0.50G22 + 0.60\ G23)/$
$(G12 + G22 + G32) \leq 0.40$ (Compound X10 in Super)

$(0.35G12 + 0.15G22 + 0.15\ G32)/$
$(G12 + G22 + G32) \leq 0.25$ (Compound X30 in Super)

The solution was found using LINDO and we noticed an alternate optimal solution.
Two solutions are found yielding a minimum cost of $1,904,000.

Decision Variable	Z = $1,940,000	Z = $1,940,000
G_{11}	0	1,400
G_{12}	0	3,500
G_{13}	14,000	9,100
G_{21}	15,000	1,100

G_{22}	7,000	20,900
G_{23}	0	0
G_{31}	0	12,500
G_{32}	25,000	7,500
G_{33}	0	4,900

Depending on whether we want to additionally minimize delivery (across different locations) or maximize sharing by having more distribution point involved, we have choices.

Example 3.7: Recruiting Raleigh Office (modified from McGrath, 2007)

Although this is a simple model, it was adopted by the US Army recruiting commend for operations. The model determines the optimal mix of prospecting strategies that a recruiter should use in a given week. The two prospecting strategies initially modeled and analyzed are phone and email prospecting. The data came from the Raleigh Recruiting Company United States Army Recruiting Command in 2006. On average, each phone lead yields 0.041 enlistments and each email lead yields 0.142 enlistments. The forty recruiters assigned to the Raleigh recruiting office prospected a combined 19,200 minutes of work per week via phone and email. The company's weekly budget is $60,000.

	Phone (x_1)	Email (x_2)
Prospecting time (Minutes)	60 minutes per lead	1 minute per lead
Budget (dollars)	$10 per lead	$37 per lead

The decision variables are:

x_1 = number of phone leads
x_2 = number of email leads

Maximize $Z = 0.041 x_1 + 0.142 x_2$

Subject to:

$60x_1 + 1x_2 \le 19,200$ (Prospecting minutes available)

$10x_1 + 37x_2 \le 60,000$ (Budget dollars available)

$x_1, x_2 \ge 0$ (non-negativity)

If we examine all the intersections point, we find a sub-optimal point, $x_1 = 294.29$, $x_2 = 154.082$, achieving 231.04 recruitments.

We examine the sensitivity analysis report.

Microsoft Excel 14.0 Sensitivity Report
Worksheet: [Book4]Sheet1
Report Created: 5/5/2015 2:21:25 PM
Variable Cells

Cell	Name	Final Value	Reduced Cost	Objective Coefficient	Allowable Increase	Allowable Decrease
$B$3	x1	294.2986425	0	0.041	8.479	0.002621622
$B$4	x2	1542.081448	0	0.142	0.0097	0.141316667

Constraints

Cell	Name	Final Value	Shadow Price	Constraint R.H. Side	Allowable Increase	Allowable Decrease
$C$10		19,200	4.38914E-05	19,200	340,579	17,518.64865
$C$11		60,000	0.003836652	60,000	648,190	56,763.16667
$C$12		294.2986425	0	1	293.2986425	1E+30
$C$13		1,542.081448	0	1	1,541.081448	1E+30

First, we see we maintain a mixed solution over a fairly large range of values for the coefficient of x_1 and x_2. Further the shadow prices provide additional information. A one unit increase in prospecting minutes available yields an increase of approximately 0.00004389 in recruits, while an increase in budget of \$1 yields an additional 0.003836652 recruits. At initial look at appears as though we might be better off with an additional \$1 in resource.

Let's assume that is cost only \$0.01 for each additional prospecting minute. Thus we could get 100*0.00004389 or a 0.004389 increase in recruits for the same unit cost increase. In this case, we would be better off obtaining the additional prospecting minutes.

3.5.1 Modeling of Ranking Units Using Data Envelopment Analysis (DEA) as an LP

Data envelopment analysis (DEA), occasionally called frontier analysis, was first put forward by Charnes, Cooper and Rhodes in 1978. It is a performance measurement technique which, as we shall see, can be used for evaluating the *relative efficiency* of *decision-making units* (*DMUs*) in organizations. Here a DMU is a distinct unit within an organization that has flexibility with respect to some of the decisions it makes, but not necessarily complete freedom with respect to these decisions.

Examples of such units to which DEA has been applied are: banks, police stations, hospitals, tax offices, prisons, defense bases (army, navy, air force), schools, and university departments. Note here that one advantage of DEA is that it can be applied to non-profit making organizations.

Since the technique was first proposed, much theoretical and empirical work has been done. Many studies have been published dealing with applying DEA in real-world situations. Obviously, there are many more unpublished studies, e.g., done internally by companies or by external consultants.

We will initially illustrate DEA by means of a small example. More about DEA can be found online using Google: "Data Envelopment Analysis". Note here that much of what you will see below is a graphical (pictorial) approach to DEA. This is very useful if you are attempting to explain DEA to those less technically qualified (such as many you might meet in the military or management world). There is a mathematical approach to DEA that can be adopted however. We will present the single measure first to demonstrate the idea and then move to multiple measures and use LP methodology from our course.

Example 3.8: Ranking Banks with DEA

Consider a number of bank branches. For each branch, we have a single output measure (number of personal transactions completed) and a single input measure (number of staff).

The data we have is as follows:

Branch	Personal transactions ('000s)	Number of staff
Branch 1	125	18
Branch 2	44	16
Branch 3	80	17
Branch 4	23	11

For example, for the branch 2 in one year, there were 44,000 transactions relating to personal accounts and 16 staff members were employed.

How then can we compare these branches and measure their performance using this data?

Ratios

A commonly used method is *ratios*. Typically we take some output measure and divide it by some input measure. Note the terminology here, we view branches as taking *inputs* and converting them (with varying degrees of efficiency, as we shall see below) into *outputs*.

For our bank branch example, we have a single input measure, the number of staff, and a single output measure, the number of personal transactions. Hence we have:

Branch	Personal transactions per staff member ('000s)
Branch 1	6.94

Branch 2	2.75
Branch 3	4.71
Branch 4	2.09

Here we can see that branch 1 has the highest ratio of personal transactions per staff member, whereas branch 4 has the lowest ratio of personal transactions per staff member.

As branch 1 has the highest ratio of 6.94, we can compare all other branches to it and calculate their *relative efficiency* with respect to branch 1. To do this, we divide the ratio for any branch by 6.94 (the value for Croydon) and multiply by 100 to convert to a percentage. This gives:

Branch .	Relative efficiency
Branch 1	$100(6.94/6.94) = 100\%$
Branch 2	$100(2.75/6.94) = 40\%$
Branch 3	$100(4.71/6.94) = 68\%$
Branch 4	$100(2.09/6.94) = 30\%$

The other branches do not compare well with branch 1, so are presumably performing less well. That is, they are relatively less efficient at using their given input resource (staff members) to produce output (number of personal transactions).

We could, if we wish, use this comparison with branch 1 to set *targets* for the other branches.

For example, we could set a target for branch 4 of continuing to process the same level of output but with one less member of staff. This is an example of an ***input target*** as it deals with an input measure.

An example of an ***output target*** would be for branch 4 to increase the number of personal transactions by 10% (e.g., by obtaining new accounts).

Plainly, in practice, we might well set a branch a mix of input and output targets which we want it to achieve. We can use LP.

3.5.2 Linear Programming Example of DEA

Example 3.9: Ranking banks with DEA as an LP

Typically, we have more than one input and one output. For the bank branch example, suppose now that we have two output measures (number of personal transactions completed and number of business transactions completed) and the same single input measure (number of staff) as before.

The data we have is as follows:

Branch	Personal transactions ('000s)	Business transactions ('000s)	Number of staff
Branch 1	125	50	18
Branch 2	44	20	16
Branch 3	80	55	17
Branch 4	23	12	11

We start be scaling (via ratios) the inputs and outputs to reflect the ratio of 1 unit.

Branch	Personal transactions ('000s)	Business transactions ('000s)	Per employee or staff
Branch 1	125/18 = 6.94	50/18 = 2.78	18/18 = 1
Branch 2	44/16 = 2.75	20/16 = 1.25	16/16 = 1
Branch 3	80/17 =4 .71	55/17 = 3.24	17/17 = 1
Branch 4	23/11 = 2.09	12/11 = 1.09	11/11 = 1

Pick a DMU to maximize: $E1$, $E2$, $E3$, or $E4$
Let $W1$ and $W2$ be the personal and business transactions at branch
In this example, we choose to maximize branch 2, $E2$.
Here is a simple LP formulation (for more on DEA, see Chapter 11):

Maximize $E2$

Subject to:
$E1 = 6.94\ W1 + 2.78\ W2$
$E2 = 2.75\ W1 + 1.25\ W2$
$E3 = 4.71\ W1 + 3.24\ W2$
$E4 = 2.09\ W1 + 1.09\ W2$
$E1 \le 1$
$E2 \le 1$
$E3 \le 1$
$E4 \le 1$

> *with (Optimization)* :

> *obj*: = *E2*;

obj: = *E2*

> *const*: = {$E1 - 6.94\ W1 - 2.78\ W2 = 0$,
    $E2 - 2.75\ W1 - 1.25\ W2 = 0, E3 - 4.71\ W1 - 3.24\ W2 = 0$,
    $E4 - 2.09\ W1 - 1.09\ W2 = 0, E1 \le 1, E2 \le 1, E3 \le 1$,
    $E4 \le 1$};

const: = {$E1 - 6.94\ W1 - 2.78\ W2 = 0, E2 - 2.75\ W1 - 1.25\ W2 = 0$,
    $E3 - 4.71\ W1 - 3.24\ W2 = 0, E4 - 2.09\ W1 - 1.09\ W2 = 0$,
    $E1 \le 1, E2 \le 1, E3 \le 1, E4 \le 1$}

> *LPSolve (obj, const, maximize)*;

[0.43149343043932,
[$E2 = 0.431493430439319481, E1 = 1.$,
$W1 = 0.0489788964841672560$,
$W2 = 0.237441172086287788, E3 = 0.99999999999999988$,
$E4 = 0.361176771225963034$]]

Now, what did we learn from this? If we ranked ordered the branches on efficiency performance of our inputs and outputs, we find

Branch 1   100%
Branch 3   100%
Branch 2   43.2%
Branch 4   36.2%

We know we need to improve on branch 2 and branch 4 performances while not losing our efficiency in branches 1 and 3. A better interpretation could be that the practices and procedures used by the other branches were to be adopted by branch 4, they could improve their performance.

This invokes issues of highlighting and disseminating examples of best practices. Equally there are issues relating to identification of poor practices.

In DEA the concept of the reference set can be used to identify best performing branches with which to compare poorly performing branches. If you use this procedure, use it wisely.

Example 3.10: Supply Chain Operations (from Fox et al., 2013)

In our case study, we present LP for supply chain design with MAPLE. We consider producing a new mixture of gasoline. We desire to minimize the total cost of manufacturing and distributing the new mixture. There is a supply chain involved with a product that must be modeled. The product is made up of components that are produced separately.

Crude Oil type	Compound A (%)	Compound B (%)	Compound C (%)	Cost/ Barrel	Barrel Avail (000 of barrels)
X10	35	25	35	$26	15,000
X20	50	30	15	$32	32,000
X30	60	20	15	$55	24,000

Demand information is as follows:

Gasoline	Compound A (%)	Compound B (%)	Compound C (%)	Expected Demand (000 of barrels)
Premium	≥ 55	≥ 23		14,000
Super		≥ 25	≤35	22,000
Regular	≥ 40		≤ 25	25,000

Let i = crude type 1, 2, 3 (X10,X20,X30, respectively)
Let j = gasoline type 1, 2, 3 (Premium, Super, Regular, respectively)

We define the following decision variables:

Gij = amount of crude i used to produce gasoline j

For example,

G_{11} = amount of crude X10 used to produce Premium gasoline
G_{12} = amount of crude type X20 used to produce Premium gasoline
G_{13} = amount of crude type X30 used to produce Premium gasoline
G_{12} = amount of crude type X10 used to produce Super gasoline
G_{22} = amount of crude type X20 used to produce Super gasoline
G_{32} = amount of crude type X30 used to produce Super gasoline
G_{13} = amount of crude type X10 used to produce Regular gasoline
G_{23} = amount of crude type X20 used to produce Regular gasoline
G_{33} = amount of crude type X30 used to produce Regular gasoline

LP formulation

$$\text{Minimize Cost} = \$86\,(G11+G21+G31)+\$92(G12+G22+G32)$$
$$+\$95(G13+G23+G33)$$

Subject to:
Demand

$G11+G21+G31 > 14,000\ (\text{Premium})$

$G12+G22+G32 > 22,000\ \left(\text{Super}\right)$

$G13+G23+G33 > 25,000\ \left(\text{Regular}\right)$

Availability of products

$G11+G12+G13 < 15,000\ \left(\text{crude 1}\right)$

$G21+G22+G23 < 32,000\ \left(\text{crude 2}\right)$

$G31+G32+G33 < 24,000\ \left(\text{crude 3}\right)$

Product mix in mixture format

$(0.35\ G11+0.50\ G21+0.60\ G31)/$
$(G11+G21+G31) \geq 0.55\ (\text{X10 in Premium})$
$(0.25\ G11+0.30\ G21+ 0.20\ G31)/$
$(G11+G21+G31) \geq 0.23\ (\text{X20 in Premium})$
$(0.35\ G13+0.15\ G23+0.15\ G33)/$
$(G13+G23+G33) \geq 0.25\ (\text{X20 in Regular})$
$(0.35\ G13+0.15\ G23+0.15\ G33)/$
$(G13+G23+G33) \leq 0.35\ (\text{X30 in Regular})$

$$(0.35 \text{ G12} + 0.50 \text{ G22} + 0.60 \text{ G23}) /$$
$$(\text{G12} + \text{G22} + \text{G32}) \leq 0.40 \ (\text{Compound X10 in Super})$$
$$(0.35 \text{ G12} + 0.15 \text{ G22} + 0.15 \text{ G32}) /$$
$$(\text{G12} + \text{G22} + \text{G32}) \leq 0.25 \ (\text{Compound X30 in Super})$$

Non-negativity

$> LPSolve\,(objgas, const, assume = nonnegative);$

$[6.02619999999905 \ 10^6, [g11 = 0., g12 = 2400.00000002064,$
$g13 = 12599.9999999794, g21 = 7000.00000001033,$
$g22 = 19599.9999999794, g23 = 0., g31 = 6999.99999998967,$
$g32 = 4599.99999998968, g33 = 12400.0000000206]]$

Decision variable	Z = $6,026,199
G_{11}	0
G_{12}	2,400.00
G_{13}	14,2599.99
G_{21}	7,000
G_{22}	19,599.9
G_{23}	0
G_{31}	6,999.99
G_{32}	4,599.99
G_{33}	12,400.00

3.5.3 Recruiting Raleigh Office*

Although this is a simple model, it was adopted by the US Army recruiting commend for operations. The model determines the optimal mix of prospecting strategies that a recruiter should use in a given week. The two prospecting strategies initially modeled and analyzed are phone and email prospecting. The data came from the Raleigh Recruiting Company United States Army Recruiting Command in 2006. On average, each phone lead yields 0.041 enlistments and each email lead yields 0.142 enlistments. The forty recruiters assigned to the Raleigh recruiting office prospected a combined 19,200 minutes of work per week via phone and email. The company's weekly budget is $60,000.

	Phone (x_1)	Email (x_2)
Prospecting time (Minutes)	60 minutes per lead	1 minute per lead
Budget (dollars)	$10 per lead	$37 per lead

---

* Modified from McGrath, 2007

The decision variables are:

x_1 = number of phone leads

x_2 = number of email leads

Maximize $Z = 0.041\ x_1 + 0.142\ x_2$

Subject to:

$60x_1 + 1x_2 \leq 19,200$ (Prospecting minutes available)

$10x_1 + 37x_2 \leq 60,000$ (Budget dollars available)

$x_1, x_2 \geq 0$ (non-negativity)

> with(Linear Algebra) : with(Optimization) : with(simplex) :

> objectivef : = 0.041 x1 + 0.1428 x2;

objectivef : = 0.041 x1 + 0.1428 x2

> const : = {60 x1 + x2 ≤ 19200, 10 x1 + 37 x2 ≤ 60000, x1 ≥ 0, x2 ≥ 0};

const : = {0 ≤ x1, 0 ≤ x2, 10 x1 + 37 x2 ≤ 60000, 60 x1 + x2 ≤ 19200}

> LPSolve(objectivef, const, maximize);

[232.275475113122, [x1 = 294.298642533937, x2 = 1542.08144796380]]

We find an optimal point, $x_1 = 294.29$, $x_2 = 152.082$, achieving 231.27 recruitments. Clearly, sensitivity analysis is important. First, we see we maintain a mixed solution over a fairly large range of values for the coefficient of x_1 and x_2. Further the shadow prices provide additional information. A one unit increase in prospecting minutes available yields an increase of approximately 0.00004389 in recruits, while an increase in budget of $1 yields an additional 0.003836652 recruits. At initial look at appears as though we might be better off with an additional $1 in resource.

Let's assume that is cost only $0.01 for each additional prospecting minute. Thus we could get 100*0.00004389 or a 0.004389 increase in recruits for the same unit cost increase. In this case, we would be better off obtaining the additional prospecting minutes.

Chapter 3 Exercises

3.1 In the supply chain case study, resolve with the following data table

Crude Oil type	Compound A (%)	Compound B (%)	Compound C (%)	Cost/ Barrel	Barrel Avail (000 of barrels)
X10	45	35	45	$26.50	18,000
X20	60	40	25	$32.85	35,000
X30	70	30	25	$55.97	26,000

Demand information is as follows:

Gasoline	Compound A (%)	Compound B (%)	Compound C (%)	Expected Demand (000 of barrels)
Premium	≥ 55	≤ 23		14,000
Super		≥ 25	≤ 35	22,000
Regular	≥ 40		≤ 25	25,000

3.2 In the Raleigh recruiting case study, assume the data has been updated as follows as resolve.

	Phone (x_1)	Email (x_2)
Prospecting time (Minutes)	45 minutes per lead	1.5 minute per lead
Budget (dollars)	$15 per lead	$42 per lead

3.3 Consider rating departments at college. The following table is provided:

DMU Departments	Inputs # Faculty	Outputs Student credit hours	Outputs Number of students	Outputs Total degrees (MS and PhD)
Unit1	25	18,341	9,086	63
Unit2	15	8,190	4,049	23
Unit3	10	2,857	1,255	31
Unit4	33	22,277	6,102	31
Unit5	12	6,830	2,910	19

Formulate and solve the DEA model and rank order the five departments.

3.4 Consider ranking companies within a Task Force. For simplification reasons, we will consider only six companies.

Companies	Inputs # Size of Unit	Output #1	Output #2	Output #3
Unit1	120	18,341	9,086	63
Unit2	110	8,190	4,049	23
Unit3	100	2,857	1,255	31
Unit4	135	22,277	6,102	31
Unit5	120	6,830	2,910	19
Unit 6	95	5,050	1,835	12

3.5 Solve the following using **graphical LP** methods and then solve using **technology** (provide the input and answer sheet from technology as a minimum). We want to maximize storage capacity of cabinets within the resource constraints of funding allocations and space available. Your supply manager has brought you the following formulated problem but needs your assistance to solve it.

a. Solve the problem graphically, state the complete answer.

Maximize $Z = 20\ x + 18\ y$ (*cubic feet of storage*)

Subject to:

$\$2,800\ x + \$2,100\ y \le \$42,000$ (*funding limit*)

$80\ x + 90\ y \le 1,440$ (*floor space in square feet*)

$x \ge 0,\ y \ge 0$

b. Solve the problem using technology, provide the answer sheet. Are there alternate solutions?

c. If you had an opportunity to procure more resources either funding or floor spaces, which would you select and why?

3.6 Solve the following using **graphical LP** methods and then solve using **technology** (provide the answer sheet as a minimum). We want to maximize storage capacity of cabinets within the resource constraints of funding allocations and space available. Your floor manager has brought you the following formulated problem but needs your assistance to solve it.
Solve the problem

Maximize $Z = 12\ x + 8\ y$ (*cubic feet of storage*)

Subject to:

$\$200\ x + \$100\ y \le \$1,400$ (*funding limit*)

$8\ x + 6\ y \le 72$ (*floor space in square feet*)

$x \ge 0,\ y \ge 0$

If I had an opportunity to procure more resources funding or floor spaces, which would I select and why?

3.7 Solve the following LP

Maximize $Z = 3\ x + 2\ y$

Subject to:

$60\ x + 40\ y \le 2,400$

$50\ x + 50\ y \le 2,500$

$x \ge 10$

$y \ge 20$

$x \ge 0,\ y \ge 0$

3.8 Jones has received the cadet loan of $18,000. Rather than buy a new car, Jones decides to invest the money in three types of stocks: low-risk, medium-risk, and high-risk. The object is to maximize the profit from these stocks. The expected returns are 7% for low-risk stocks, 10% for medium-risk stocks, and 15% of high-risk stocks. Jones is not very cautious and will invest at most $2,000 more in low-risk than in medium risk, at most $8,000 in high-risk stocks, and no more than $14,000 in both medium-risk and high-risk stocks. (a) Identify the decision variables for this problem. (b) Formulate the problem. DO NOT SOLVE! (c) List several of the assumptions required for LP. Is the above problem an LP problem?

3.9 Given:

Maximize $Z = 10x_1 + 20x_2$

Subject to:

$$-x_1 + 2x_2 \leq 25$$
$$x_1 + x_2 \leq 22$$
$$5x_1 + 3x_2 \leq 55$$
$$x_1 \geq 0$$
$$x_2 \geq 0$$

a. Solve the problem graphically. (b) Show that the sufficient condition holds at optimality, i.e., show in an accurate graph the gradient of the objective function lies in the cone spanned by all the binding constraints. (c) Define "binding constraints".

3.10 Given:

Maximize $Z = 10x_1 + 20x_2$

Subject to:

$$-x_1 + 2x_2 \leq 15$$
$$x_1 + x_2 \leq 12$$
$$5x_1 + 3x_2 \leq 45$$
$$x_1 \geq 0$$
$$x_2 \geq 0$$

Solve via technology.

3.11 Formulation: The Marvel Toy Company wishes to make three models of boats for the most profit. They found that a model of a steamboat takes the cutter 1 hour, the painter 2 hours, and the assembler 4 hours. It produces $6 profit. The model of a 4-mast sailboat takes the cutter 3 hours, the painter 3 hours, and the assembler 2.5 hours. It has a $3.50 profit. The clipper ship takes the cutter 1.5 hours, the painter 2.5 hours, and the assembler 1 hour. It provides a $2.75 profit.

The cutter is available for only 45 hours, the painter for 50 hours, and the assembler for 60 hours. Assume they can sell all that they build. (a) Identify the decision variables for this problem. (b) Formulate the problem. DO NOT SOLVE! (c) List several of the assumptions required for LP. Is the above problem an LP problem?

3.12 A manufacturer of an industrial product has to meet the following shipping schedule.

Month	Required Shipment (Units)
January	10,000
February	40,000
March	20,000

The monthly production capacity is 30,000 units and the production cost per unit is $10. Since the company does not own their own warehouse, the service of a storage company is utilized whenever needed. The storage company figures its bill by multiplying the number of units in storage on the last day of the month by $3. On the first day of January, the company does not have any beginning inventory, and it doesn't want to have any ending inventory at the end of March. Formulate a model to assist in minimizing the sum of production and storage costs for the three month period.

3.13 Solve the following LP where the decision variables are:

$X1$ = number of desks

$X2$ = number of tables

$X3$ = number of chairs

Maximize $Z = 60\ X1 + 30\ X2 + 20\ X3$

Subject to:

$8\ X1 + 6\ X2 + X3 <= 48$ (lumber)

$4\ X1 + 2\ X2 + 1.5\ X3 <= 20$ (finishing hours)

$2\ X1 + 1.5\ X2 + 0.5\ X3 <= 8$ (carpentry hours)

$X2 <= 5$ (limitation on demand)

$X1, X2, X3 \geq 0$

3.14 Solve the following LP:

Minimize $Z = 15\ x + 20\ y$

Subject to:

$x + 2\ y \geq 10$

$2\ x - 3\ y \leq 6$

$x + y \geq 6$

$x \geq 0$

$y \geq 0$

3.15 Solve the following LP:

Minimize $Z = 15\,x + 20\,y$

Subject to:

$x + 2\,y \geq 10$

$x + y \geq 6$

$x \geq 0$

$y \geq 0$

3.16 Solve the following LP:

Minimize $Z = 10\,x + 7\,y$

Subject to:

$3\,x + 2\,y \geq 6$

$2\,x + 4\,y \geq 8$

$x, y \geq 0$

3.17 Solve the following:

Maximize $P = 140\,x + 120\,y$

Subject to:

$2\,x + 4\,y <= 1,400$

$4x + 3\,y <= 1,500$

$x, y \geq 0$

3.18 Maximize $f(x_1, x_2) = 3\,x_1 + 2x_2$

Subject to:

$6x_1 + 4x_2 \leq 24$

$10x_1 + 3x_2 \leq 30$

$x_1, x_2 \geq 0$

Projects

In the following projects, complete these requirements.
Required:

a. List the decision variables and define them.

b. List the objective function.

c. List the resources that constrain this problem.

d. Graph the "feasible region".

e. Label all intersection points of the feasible region.

f. Plot the Objective function in a different color (highlight the Objective function line, if necessary) and label it the ISO-Profit line.

g. Clearly indicate on the graph the point that is the optimal solution.

h. List the coordinates of the optimal solution and the value of the objective function.

i. Assume now that manufacturer of additives has the opportunity to sell you a nice TV special deal to deliver at least .5 lb. of Additive 1 and at least .3 lbs. of Additive 2. Use graphical LP methods to help recommend whether you should buy this TV offer. Support your recommendation.

j. Write a one-page cover letter to your boss of the company that summarizes the results that you found.

3.1 With the rising cost of gasoline and increasing prices to consumers, the use of additives to enhance performance of gasoline is being considered. Consider two additives, Additive 1 and Additive 2. The following conditions must hold for the use of additives:

- Harmful carburetor deposits must not exceed 1/2 lb. per car's gasoline tank.

- The quantity of Additive 2 plus twice the quantity of Additive 1 must be at least 1/2 lb. per car's gasoline tank.

- 1 lb. of Additive 1 will add 10 octane units per tank, and 1 lb. of Additive 2 will add 20 octane units per tank. The total number of octane units added must not be less than six (6).

- Additives are expensive and cost $1.53/lb. for Additive 1 and $4.00/lb. for Additive 2.

  We want to determine the quantity of each additive that will meet the above restrictions and will minimize their cost.

3.2 A farmer has 30 acres on which to grow tomatoes and corn. Each 100 bushels of tomatoes require 1,000 gallons of water and 5 acres of land. Each 100 bushels of corn require 6,000 gallons of water and 2 1/2 acres of land. Labor costs are $1 per bushel for both corn and tomatoes. The farmer has available 30,000 gallons of water and $750 in capital. He knows that he cannot sell more than 500 bushels of tomatoes or 475 bushels of corn. He estimates a profit of $2 on each bushel of tomatoes and $3 of each bushel of corn. How many bushels of each should he raise to maximize profits?

3.3 *Fire Stone Tires* headquartered in Akron, OH has a plant in Florence, SC which manufactures two types of tires: SUV 225 Radials and

SUV 205 Radials. Demand is high because of the recent recall of tires. Each 100 SUV 225 Radials requires 100 gallons of synthetic plastic and 5 lbs. of rubber. Each 100 SUV 205 Radials require 60 gallons synthetic plastic and 2 1/2 lbs. of rubber. Labor costs are $1 per tire for each type tire. The manufacturer has weekly quantities available of 660 gallons of synthetic plastic, $750 in capital, and 300 lbs. of rubber. The company estimates a profit of $3 on each SUV 225 radial and $2 of each SUV 205 radial. How many of each type tire should the company manufacture in order to maximize their profits?

3.4 Consider a toy maker that carves wooden soldiers. The company specializes in two types: Confederate soldiers and Union soldiers. The estimated profit for each is $28 and $30, respectively. A Confederate soldier requires 2 units of lumber, 4 hours of carpentry, and 2 hours of finishing to complete the soldier. A Union soldier requires 3 units of lumber, 3.5 hours of carpentry, and 3 hours of finishing to complete. Each week the company has 100 units of lumber delivered. The workers can provide at most 120 hours of carpentry and 90 hours of finishing. Determine the number of each type wooden soldiers to produce to maximize weekly profits. Formulate and then solve this LP graphically.

3.5 A company wants to bottle 2 different drinks for the holiday season. It takes 3 hours to bottle one gross of Drink A, and it takes 2 hour to label the bottles. It takes 2.5 hours to bottle one gross of Drink B, and it takes 2.5 hours to label the bottles. The company makes $15 profit on one gross of Drink A and an $18 profit of one gross of Drink B. Given that we have 40 hours to devote to bottling the drinks and 35 hours to devote to labeling bottles per week, how many bottles of each type drink should the company package to maximize profits?

3.6 The Mariners Toy Company wishes to make three models of ships to maximize their profits. They found that a model steamship takes the cutter 1 hour, the painter 2 hours, and the assembler 4 hours of work; it produces a profit of $6.00. The sailboat takes the cutter 3 hours, the painter 3 hours, and the assembler 2 hours. It produces a $3.00 profit. The submarine takes the cutter 1 hour, the painter 3 hours, and the assembler 1 hour. It produces a profit of $2.00. The cutter is only available for 45 hours per week, the painter for 50 hours, and the assembler for 60 hours. Assume that they sell all the ships that they make, formulate this LP to determine how many ships of each type that Mariners should produce.

3.7 In order to produce 1,000 tons of non-oxidizing steel for BMW engine valves, at least the following units of manganese, chromium, and molybdenum, will be needed weekly: 10 units of manganese, 12 units of chromium, and 14 units of molybdenum (1 unit is 10 lbs.). These materials are obtained from a dealer who markets these metals in three sizes small (S), medium (M), and large (L). One S case costs $9 and contains 2 units of manganese, 2 units of chromium, and 1 unit of molybdenum. One M case costs $12 and contains 2 units of manganese, 3 units of chromium, and 1 unit of molybdenum. One L case costs $15 and contains 1 unit of manganese, 1 units of chromium, and 5 units of molybdenum. How many cases of each kind (S, M, L) should be purchased weekly so that we have enough manganese, chromium, and molybdenum at the smallest cost?

3.8 The Super bowl Advertising agency wishes to plan an advertising campaign in three different media – television, radio, and magazines. The purpose or goal is to reach as many potential customers as possible. Results of a marketing study are given below:

	Day time TV	Prime Time, TV	Radio	Magazines
Cost of advertising Unit	$40,000	$75,000	$30,000	$15,000
Number of potential customers reached per unit	400,000	900,000	500,000	200,000
Number of woman customers reached per unit	300,000	400,000	200,000	100,000

The company does not want to spend more than $800,000 on advertising. It further requires (1) at least 2 million exposures take place among woman; (2) TV advertising be limited to $500,000; (3) at least three advertising units be bought on day time TV and two units on prime time TV, and (4) the number of radio and magazine advertisement units should each be between 5 and 10 units.

3.9 A tomato cannery has 5,000 pounds of grade A tomatoes and 10,000 pounds of grade B tomatoes, from which they will make whole canned tomatoes and tomato paste. Whole tomatoes must be composed of at least 80% grade A tomatoes, whereas tomato paste must be made with at least 10% grade A tomatoes. Whole tomatoes well for $0.08 per pound and grade B tomatoes sell for $0.05 per pound. Maximize revenue of the tomatoes.

HINT: Let X_{wa} = pounds of grade A tomatoes used to whole tomatoes, X_{wb} = pounds of grade B tomatoes used to whole tomatoes. The total number of whole tomato cans produced is the sum of $X_{wa} + X_{wb}$

after each is found. Also remember a percent is a fraction of the whole times 100%.

3.10 The McCow Butchers is a large-scale distributor of dressed meats for Myrtle Beach restaurants and hotels. Ryan's order meat for meat-loaf (mixed ground beef, pork, and veal) for 1,000 lbs. according to the following specifications:

a. Ground beef is to be no less than 400 lbs. and no more than 600 lbs.

b. The ground pork is to between 200 and 300 lbs.

c. The ground veal must weigh between 100 and 400 lbs.

d. The weight of the ground pork must be no more than one and one half (3/2) times the weight of the veal.

The contract calls for Ryan's to pay $1,200 for the meat. The cost per pound for the meat is: $0.70 for hamburger, $0.60 for pork, and $0.80 for the veal. How can this be modeled?

3.11 **Portfolio Investments**

A portfolio manager in charge of a bank wants to invest $10 million. The securities available for purchase, as well as their respective quality ratings, maturate, and yields, are shown in Table 3.5.

TABLE 3.5

Managers Data

Bond Name	Bond Type	Moody's Quality Scale	Bank's Quality Scale	Years to Maturity	Yield at Maturity	After-Tax Yield
A	MUNICI-PAL	Aa	2	9	4.3%	4.3%
B	AGENCY	Aa	2	15	5.4%	2.7%
C	GOVT 1	Aaa	1	4	5%	2.5%
D	GOVT 2	Aaa	1	3	4.4%	2.2%
E	LOCAL	Ba	5	2	4.5%	4.5%

The bank places certain policy limitations on the portfolios manager's actions:

a. Government and Agency Bonds must total at least $4 million.

b. The average quality of the portfolios cannot exceed 1.4 on the Bank's quality scale. Note a low number means high quality.

c. The average years to maturity must not exceed 5 years.

Assume the objective is to maximize after-tax earnings on the investment.

References

Fox, W. P. (2013). Book Chapter, "http://www.intechopen.com/books/engineer-ing-management/modeling-engineering-management-decisions-with-game-theory" **Modeling Engineering Management Decisions with Game Theory**, "http://www.intechopen.com/books/engineering-management" "Engineering Management" edited by Fausto Pedro García Márquez and Benjamin Lev, ISBN 978-953-51-1037-8, InTech, March 3, 2013.

McGrath, G. (2007). Email marketing for the U.S. Army and Special Operations Forces Recruiting. Master's Thesis, Naval Postgraduate School, Monterey, CA.

Additional Readings

Apaiah, R. & E. Hendrix. (2006). Linear programming for supply chain design: A case on Novel protein foods. PhD Thesis, Wageningen University, The Netherlands.

Balakrishnan, N., B. Render, & R. Stair. (2007). *Managerial Decision Making*, 2nd ed. Saddle River, NJ: Prentice Hall.

Bazarra, M.S., J.J. Jarvis, & H.D. Sheralli. (1990). *Linear Programming and Network Flows.* New York, NY: John Wiley & Sons.

Ecker, J. & M. Kupperschmid. (1988). *Introduction to Operations Research.* New York, NY: John Wiley and Sons.

Fox, W. (2013). *Mathematical Modeling with Maple.* Boston, MA: Cengage Publishers.

Fox, W. (2018). *Mathematical Modeling for Business Analytics.* Boca Raton, FL: CRC Press.

Giordano, F., W. Fox, & S. Horton. (2013). *A First Course in Mathematical Modeling*, 5th ed. Boston, MA: Cengage.

Hiller, F.S. & G.J. Liberman. (1990). *Introduction to Mathematical Programming.* New York, NY: McGraw Hill Publishing Company.

Winston, W.L. (2002). *Introduction to Mathematical Programming Applications and Algorithms*, 4th ed. Belmont, CA: Duxbury Press.

Winston, W. (2005). *Mathematical Programming.* Boston, MA: Cengage.

4

Multi-Attribute Decision-Making Using Weighting Schemes with SAW, AHP, and TOPSIS

4.1 Weighting Methods

Weighting methods are critical to the multi-attribute decision process. We will discuss four methods: rank order centroid method (ROC), ratio method, pairwise comparisons of analytical hierarchy process (AHP) method, and entropy method.

4.1.1 Rank Order Centroid (ROC)

This method is a simple way of giving weight to a number of items ranked according to their importance. The decision makers usually can rank items much more easily than give weight to them. This method takes those ranks as inputs and converts them to weights for each of the items. The conversion is based on the following formula:

$$w_i = \left(\frac{1}{M}\right)\sum_{n=i}^{M}\frac{1}{n}$$

1. List objectives in order from most important to least important
2. Use the above formulas for assigning weights

where M is the number of items, and W_i is the weight for the i item. For example, if there are four items, the item ranked first will be weighted $(1 + 1/2 + 1/3 + 1/4)/4 = 0.52$, the second will be weighted $(1/2 + 1/3 + 1/4)/4 = 0.27$, the third $(1/3 + 1/4)/4 = 0.15$, and the last $(1/4)/4 = 0.06$. As shown in this example, the ROC is simple and easy to follow, but as shown it gives weights which are highly dispersed. As an example, consider the same factors to be weighted (shortening schedule, agency control over the project, project cost, and competition). If they are ranked based on their importance

DOI: 10.1201/9781032726885-4

and influence on decision as 1 – shortening schedule, 2 – project cost, 3 – agency control over the project, and 4 – competition, their weights would be 0.52, 0.27, 0.15, and 0.06, respectively. These weights almost eliminate the effect of the fourth factor, i.e., among competitors. This could be an issue for a decision maker.

4.1.2 Ratio Method for Weights

The ratio method is another simple way of calculating weights for a number of critical factors. A decision maker should first rank all the items according to their importance. The next step is giving weight to each item based on its rank. The lowest ranked item will be given a weight of 10. The weight of the rest of the items should be assigned as multiples of 10. The last step is normalizing these raw weights. This process is shown in the example below. Note that the weights should not necessarily jump 10 points from one item to the next. Any increase in the weight is based on the subjective judgment of the decision maker and reflects the difference between the importance of the items. Ranking the items in the first step helps in assigning more accurate weights (Table 4.1).

Normalized weights are simply calculated by dividing the raw weight of each item over the sum of the weights for all items. For example, normalized weight for the first item (shortening schedule) is calculated as $50/(50 + 40 + 20 + 10) = 41.7\%$. The sum of normalized weights is equal to 100% ($41.7 + 33.3 + 16.7 + 8.3 = 100$).

4.1.3 Pairwise Comparison (AHP)

In this method, the decision maker should compare each item with the rest of the group and give a preferential level to the item in each pairwise comparison (Chang, 2004). For example, if the item at hand is as important as the second one, the preferential level would be 1. If it is much more important, its level would be 10. After conducting all of the comparisons and determining the preferential levels, the numbers will be added up and normalized. The results are the weights for each item (Table 4.2). Table 4.3 can be used as a guide for giving a preferential level score to an item while comparing it with another one. The following example shows the application of the pairwise

TABLE 4.1

Ratio Method

Task/Item	Shorten Schedule	Project Cost	Agency Control	Competition
Ranking	1	2	3	4
Weighting	50	40	20	10
Normalizing	41.7%	33.3%	16.7%	8.3%

TABLE 4.2

Saaty's Nine-Point Scale

Intensity of Importance in Pairwise Comparisons	Definition
1	Equal importance
3	Moderate importance
5	Strong importance
7	Very strong importance
9	Extreme Importance
2,4,6,8	For comparing between the above
Reciprocals of above	In comparison of elements *i* and *j* if *i* is 3 compared to *j*, then j is 1/3 compared to *i*.

comparison procedure. Referring to the four critical factors identified above, let us assume that shortening the schedule, project cost, and agency control of the project are the most important parameters in the project delivery selection decision. Following the pairwise comparison, the decision maker should pick one of these factors (e.g., shortening the schedule) and compare it with the remaining factors and give a preferential level to it. For example, shortening the schedule is more important than project cost; in this case, it will be given a level of importance of the 5. Pairwise comparisons use the information in Table 4.3.

The decision maker should continue the pairwise comparison and give weights to each factor. The weights, which are based on the preferential levels given in each pairwise comparison, should be consistent to the extent possible. The consistency is measured based on the matrix of preferential levels. The interested reader can find the methods and applications of consistency measurement in Temesi (2006).

TABLE 4.3

Pairwise Comparison Example

	Shorten the Schedule (1)	Project Cost (2)	Agency Control (3)	Competition (4)	Total (5)	Weights (6)
Shorten the schedule	1	5	5/2	8	16.5	16.5 / 27.225 = 0.60
Project cost	1/5	1	½	1	2.7	2.7 / 27 / 225 = 0.10
Agency control	2/5	2	1	2	5.4	5.4 / 27 / 225 = 0.20
Competition	1/8	1	½	1	2.625	2.625 / 27 / 225 = 0.10
					Total = 27.225	1

Table 4.3 shows the rest of the hypothetical weights and the normalizing process, the last step in the pairwise comparison approach.

Note that Column 5 is simply the sum of the values in Columns 1–4. Also note that if the preferential level of factor i to factor j is n, then the preferential level of factor j to factor i is simply $1/n$. The weights calculated for this exercise are 0.6, 0.1, 0.2, and 0.1 which add up to 1.0. Note that it is possible for two factors to have the same importance and weight.

4.1.4 Entropy Method

Shannon and Weaver (1948) proposed the entropy concept and this concept has been highlighted by Zeleny et al. (1982) for deciding the weights of attributes. Entropy is the measure of uncertainty in the information using probability methods. It indicates that a broad distribution represents more uncertainty than a sharply peaked distribution.

To determine the weights by the entropy method, the normalized decision matrix we call R_{ij} is considered. The equation used is

$$e_j = -k \sum_{i=1}^{n} R_{ij} \ln\left(R_{ij}\right)$$

where $k = 1/ln(n)$ is a constant that guarantees that $0 \le e_j \le 1$. The value of n refers to the number of alternatives. The degree of divergence (d_j) of the average information contained by each attribute can be calculated as:

$$d_j = 1 - e_j.$$

The more divergent the performance rating R_{ij}, for all i and j, then the higher the corresponding d_j and the more important the attribute B_j is considered to be.

The weights are found by the equation, $w_j = \frac{(1-e_j)}{\Sigma(1-e_j)}$.
Let's do an example to obtain entropy weights.

Example 4.1: Entropy for Cars

a. The data:

	Cost	Safety	Reliability	Performa	MPG City	MPG HW	Interior/ Style
a1	27.8	9.4	3	7.5	44	40	8.7
a2	28.5	9.6	4	8.4	47	47	8.1
a3	38.668	9.6	3	8.2	35	40	6.3
a4	25.5	9.4	5	7.8	43	39	7.5
a5	27.5	9.6	5	7.6	36	40	8.3
a6	36.2	9.4	3	8.1	40	40	8

b. Sum the columns

sums	184.168	57	23	47.6	245	246	46.9

c. Normalize the data. Divide each data element in a column by the sum of the column.

0.150949	0.164912	0.13043478	0.157563	0.17959184	0.162602	0.185501066
0.15475	0.168421	0.17391304	0.176471	0.19183673	0.191057	0.172707889
0.20996	0.168421	0.13043478	0.172269	0.14285714	0.162602	0.134328358
0.138461	0.164912	0.2173913	0.163866	0.1755102	0.158537	0.159914712
0.14932	0.168421	0.2173913	0.159664	0.14693878	0.162602	0.176972281
0.19656	0.164912	0.13043478	0.170168	0.16326531	0.162602	0.170575693

d. Use the entropy formula, where in the case $k = 6$.

$$e_j = -k \sum_{i=1}^{n} R_{ij} \ln\left(R_{ij}\right)$$

e1	e2	e3	e4	e5	e6	e7	k = 0.558111
−0.28542	−0.29723	−0.2656803	−0.29117	−0.3083715	−0.29536	−0.31251265	
−0.28875	−0.30001	−0.3042087	−0.30611	−0.31674367	−0.31623	−0.30330158	
−0.32771	−0.30001	−0.2656803	−0.30297	−0.27798716	−0.29536	−0.26965989	
−0.27376	−0.29723	−0.3317514	−0.29639	−0.30539795	−0.2919−	−0.293142	
−0.28396	−0.30001	−0.3317514	−0.29293	−0.28179026	−0.29536	−0.3064739	
−0.31976	−0.29723	−0.2656803	−0.30136	−0.29589857	−0.29536	−0.3016761	

e. Find *ej*,

0.993081	0.999969	0.98492694	0.999532	0.99689113	0.998825	0.997213162

f. Compute weights by formula,

	0.006919	3.09E-05	0.01507306	0.000468	0.00310887	0.001175	0.002786838	0.029561
w	0.234044	0.001046	0.50989363	0.015834	0.1051674	0.039742	0.094273533	

g. Check that weights sum to 1, as they did above.
h. Interpret weights and rankings.
i. Use these weights in further analysis.

Let's see the possible weights under each method.

1		Safety	A	2
2		Reliability	A	3
3		Performance	A	3
4	Cost versus	MPG City	A	4
5		MPG HW	A	5
6		Interior/Style	A	7
7				
1		Reliability	A	2
2		Performance	A	2
3	Safety vs	MPG City	A	3
		MPG HW		
4			A	4
5	Reliability	Performace	A	2
6		MPG City	A	3
7	Performace	MPG City	A	2
		MPG HW		
2			A	4
3	MPG City	MPG HW	A	3
4				
	MPG HW vs	Interior/Styl e	A	3

Results were:

	c1
Cost	0.36123313
Safety	0.2093244
Reliability	0.14458999
Performance	0.11667294
MPG City	0.0801478
MPG HW	00.05298706
Interior/Style	00.03504467

If we were to have used the ratio method, these would be our weights
Ratio Method:

Criteria in Order	Cost	Safety	Reliability	Performance	MPG City	MPG HW	Interior/ Style	
Values	70	60	50	40	30	20	10	280
Weights	0.25	0.214	0.179	0.143	0.107	0.714	0.358	Sums to 1

4.2 Simple Additive Weights (SAW) Method

4.2.1 Description and Uses

This is also called the weighted sum method (Fishburn, 1967) and is the simplest, and still one of the widest, used Multi-Attribute Decision Making (MADM) methods. Its simplistic approach makes it easy to use. Depending on the type of relational data used, we might either want the larger average or the smaller average.

4.2.2 Methodology

Here, each criterion (attribute) is given a weight, and the sum of all weights must be equal to 1. Each alternative is assessed with regard to every criterion (attribute). The overall or composite performance score of an alternative is given simply by Equation (4.1) with m criteria.

$$P_i = \left(\sum_{j=1}^{m} w_j m_{ij} \right) / m \qquad (4.1)$$

Previously, it was argued that simple additive weights (SAW) should be used only when the decision criteria can be expressed in identical units of measure (e.g., only dollars, only pounds, only seconds, etc.). However, if all the elements of the decision table are normalized, then this procedure can be used for any type and number of criteria. In that case, Equation (4.1) will take the following form still with m criteria shown as Equation (4.2):

$$P_i = \left(\sum_{j=1}^{m} w_j m_{ijNormalized} \right) / m \qquad (4.2)$$

where $(m_{ijNormalized})$ represents the normalized value of m_{ij}, and P_i is the overall or composite score of the alternative A_i. The alternative with the highest value of P_i is considered the best alternative.

4.2.3 Strengths and Limitations

The strengths are the ease of use and the normalized data allow for comparison across many differing criteria. Limitations include larger is always better or smaller is always better. There is not the flexibility in this method to state which criterion should be larger or smaller to achieve better performance. This makes gathering useful data of the same relational value scheme (larger or smaller) essential.

4.2.4 Sensitivity Analysis

Sensitivity analysis should be applied to the weighting scheme employed to determine how sensitive the model is to the weights. Weighting can be arbitrary for a decision maker, or in order to obtain weights, one might choose to use a scheme to perform pairwise comparison, as shown in AHP, which we discuss later. Whenever subjectivity enters into the process for finding weights, then sensitivity analysis is recommended. Please see later sections for a suggested scheme for dealing with sensitivity analysis for individual criteria weights.

4.2.5 Illustrative Examples Using SAW

Example 4.2: Car Selection using SAW (Data from Consumer's Report and US News and World Report Online Data)

We are considering six cars: Ford Fusion, Toyota Prius, Toyota Camry, Nissan Leaf, Chevy Volt, and Hyundai Sonata. For each car, we have data on seven criteria that were extracted from Consumer's Report and US News and World Report data sources. They are *cost, mpg city, mpg highway, performance, interior and style, safety,* and *reliability.* We provide the extracted information in Table 4.4:

TABLE 4.4

Raw Data

Cars	Cost ($000)	MPG City	MPG Hw	Performance	Interior and Style	Safety	Reliability
Prius	27.8	44	40	7.5	8.7	9.4	3
Fusion	28.5	47	47	8.4	8.1	9.6	4
Volt	38.668	35	40	8.2	6.3	9.6	3
Camry	25.5	43	39	7.8	7.5	9.4	5
Sonata	27.5	36	40	7.6	8.3	9.6	5
Leaf	36.2	40	40	8.1	8.0	9.4	3

Initially, we might assume all weights are equal to obtain a baseline ranking. We substitute the rank orders (1st–6th) for the actual data. We compute the average rank attempting to find the best ranking (smaller is better). We find that our rank ordering is Fusion, Sonata, Camry, Prius, Volt, and Leaf.

SAW Using Rank Ordering of Data by Criteria

	cost	MPG City	MPG HW	Perf	Interior style	Safety	Reliability		Rank	
Prius	3	2	2	6	1	2	4		2.857143	4
Fusion	4	1	1	1	3	1	3		2	1
volt	6	6	2	2	6	1	4		3.857143	6
Camry	1	3	3	4	5	2	1		2.714286	3
Sonata	2	5	2	5	2	1	1		2.571429	2
Leaf	5	4	2	2	4	2	4		3.285714	5

Next, we apply a scheme to the weights and still use the ranks 1–6 as before. Perhaps we apply a technique similar to the pairwise comparison that we will discuss in the AHP, Section 4. Using the pairwise comparison to obtain new weights, we obtain a new ordering: Camry, Sonata, Fusion, Prius, Leaf, and Volt. The changes in results of the rank ordering differ from using equal weights and shows the sensitivity that the model has to the given criteria weights. We assume the criteria in order of importance are: cost, reliability, MPG City, safety, MPG HW, performance, interior and style.

We use pairwise comparisons to obtain weights:

Unequal Weights

weights	0.311156	0.133614	0.095786	0.055069	0.049997069	0.129372	0.225007
	cost	MPG City	MPG HW	Perf	Interior style	Safety	Reliability

Using these weights and applying them to the previous ranking, we obtain values that we average, and then we select the smaller average.

Cars	Weighted Cost	Weighted MPG City	Weighted MPG HW	Weighted Perf	Weighted Interior st	Weighted Safety	Reliability	Average	Ranking
Prius	0.933468	0.267228	0.191572	0.330411637	0.049997	0.258743	17.77726	0.33857	4
Fusion	1.244624	0.133614	0.095786	0.055068606	0.149991	0.129372	13.33294	0.301409	2
volt	1.866936	0.801684	0.191572	0.110137212	0.299982	0.129372	17.77726	0.566614	6
Camry	0.311156	0.400842	0.287359	0.220274424	0.249985	0.258743	4.444315	0.28806	1
Sonata	0.622312	0.66807	0.191572	0.275343031	0.099994	0.129372	4.444315	0.331111	3
Leaf	1.55578	0.534456	0.191572	0.110137212	0.199988	0.258743	17.77726	0.475113	5

SAW Using Raw Data

We could use the raw data directly from Table 4.2 except cost. Now, only *cost* represents a value where smaller is better so we can replace cost with its reciprocal. So 1/*cost* represents a variable where larger is better. If we use the criteria weights from the previous results and our raw data replacing *cost* with 1/*cost*, we obtain a final ranking based upon larger values are better: Fusion (1.972), Camry (1.805), Prius (1.780), Leaf (1.703), Sonata (1.693), Volt (1.599).

Sensitivity Analysis

We suggest employing sensitivity analysis on the criteria weights that are used in the model as presented and described by varying subjective weights.

Example 4.3: Kite Network

We visited the Kite Network described earlier. Here we present two methods that will work on the data from Example 2 from the previous section. Method I represents transforming the output data into rankings from 1st to last place. Then we apply the weights and average all the values. We rank them from smaller to larger to represent the alternative choices. We present only results using the pairwise compare criteria to obtain the weighted criteria.

Weights	0.153209	0.144982	0.11944	0.067199	0.157688	0.357482
Susan	1	1	10	1	3	2
Steve	2	2	2	2	1	4
Sarah	3	2	1	10	1	7
Tom	4	4	2	2	5	5
Claire	4	4	2	2	5	5
Fred	6	7	2	2	7	8
David	6	7	2	2	7	8
Claudia	6	6	2	2	3	1
Ben	9	9	2	2	9	3
Jennifer	10	10	2	2	10	8

Susan	0.153209	0.144982	1.194396	0.067199	0.473064	0.714965	0.457969	Steve	0.426213
Steve	0.306418	0.289964	0.238879	0.134398	0.157688	1.42993	0.426213	Susan	0.457969
Sarah	0.459627	0.289964	0.11944	0.67199	0.157688	2.502377	0.700181	Claudia	0.498828
Tom	0.612835	0.579928	0.238879	0.134398	0.78844	1.787412	0.690316	Tom	0.690316
Claire	0.612835	0.579928	0.238879	0.134398	0.78844	1.787412	0.690316	Claire	0.690316
Fred	0.919253	1.014875	0.238879	0.134398	1.103816	2.859859	1.04518	Sarah	0.700181
David	0.919253	1.014875	0.238879	0.134398	1.103816	2.859859	1.04518	Ben	0.924772
Claudia	0.919253	0.869893	0.238879	0.134398	0.473064	0.357482	0.498828	Fred	1.04518
Ben	1.37888	1.304839	0.238879	0.134398	1.419192	1.072447	0.924772	David	1.04518
Jennifer	1.532089	1.449821	0.238879	0.134398	1.576879	2.859859	1.298654	Jennifer	1.298654

Method I rankings: Steve, Susan, Claudia. Tom, Claire, Sarah, Ben Fred, David, and Jennifer.

Method II uses the raw metrics data and the weights as above where larger values are better.

Susan	0.027663	0.025384	0.010988	0.007262	0.017164	0.0723	0.026793	Claudia	0.029541
Steve	0.021279	0.019939	0.011903	0.006743	0.017833	0.055503	0.0222	Susan	0.026793
Sarah	0.019151	0.019939	0.013226	0.005994	0.017833	0.037245	0.018898	Steve	0.0222
Tom	0.017023	0.016586	0.011903	0.006743	0.015916	0.006938	0.012518	Sarah	0.018898
Claire	0.017023	0.016586	0.011903	0.006743	0.015916	0.006938	0.012518	Ben	0.018268
Fred	0.012767	0.013593	0.011903	0.006743	0.015381	0	0.010065	Tom	0.012518
David	0.012767	0.013593	0.011903	0.006743	0.015381	0	0.010065	Claire	0.012518
Claudia	0.012767	0.015108	0.011903	0.006743	0.017164	0.113562	0.029541	Fred	0.010065
Ben	0.008512	0.003497	0.011903	0.006743	0.013954	0.064997	0.018268	David	0.010065
Jennifer	0.004256	0.000757	0.011903	0.006743	0.011146	0	0.005801	Jennifer	0.005801

The results are Claudia, Susan, Steven, Sarah, Ben, Tom, Claire, Fred, David, and Jennifer. Although the top three are the same their order is different. The model is sensitive both to the input format and the weights.

Sensitivity Analysis

We can apply sensitivity analysis to the weights, in a controlled manner, and determine each change's impact on the final rankings.

We used Equation (4.3) (Alinezhad and Amini, 2011) for adjusting weights:

$$w'_j = \frac{1 - w'_p}{1 - w_p} w_j \qquad (4.3)$$

where w_j' is the new weight, w_p is the original weight of the criterion to be adjusted, and w_p' is the value after the criterion was adjusted. We found this to be an easy method to adjust weights to reenter back into our model.

4.3 Weighted Product Method

This method is very similar to SAW. The main difference is that, instead of addition in the model, there is multiplication (Miller and Starr, 1969). The overall composite performance score of each alternative is given by

$$P_i = \prod_{j=1}^{m} \left[(m_{ij}) normalized \right]^{w_j}$$

The normalized values are calculated as explained under SAW. The alternative with the highest Pi value is best.

Example 4.4: Weighted Product Method

This simple decision problem is based on three alternatives denoted as $A1$, $A2$, and $A3$ each described in terms of four criteria $C1$, $C2$, $C3$, and $C4$. Next, let the numerical data for this problem be as in the following decision matrix:

	C1	C2	C3	C4
Alts.	0.20	0.15	0.40	0.25
A1	25	20	15	30
A2	10	30	20	30
A3	30	10	30	10

From the above data, we can easily see that the relative weight of the first criterion is equal to 0.20, the relative weight for the second criterion is 0.15, and so on. Similarly, the value of the first alternative (i.e., $A1$) in terms of the first criterion is equal to 25, the value of the same alternative in terms of the second criterion is equal to 20, and so on. However, now the restriction to express all criteria in terms of the same measurement unit is not needed. That is, the numbers under each criterion may be expressed in different units.

Product Method

	C1	C2	C3	C4
Alts.	0.2	0.15	0.4	0.25
A1	25	20	15	30
A2	10	30	20	30
A3	30	10	30	10

sums	65	60	65	70		
Normalize						
	0.384615	0.333333	0.230769	0.428571		
	0.153846	0.5	0.307692	0.428571		
	0.461538	0.166667	0.461538	0.142857		
						ranks
	0.826048	0.84807	0.556251	0.809107	0.315293	1
	0.687729	0.90125	0.624089	0.809107	0.312979	2
	0.856725	0.764324	0.733978	0.614788	0.29548	3

Therefore, the best alternative is $A1$, since it is superior to all the other alternatives.

4.4 Analytical Hierarchy Process (AHP)

4.4.1 Description and Uses

AHP is a multi-objective decision analysis tool first proposed by Saaty (1980). It is designed when either subjective and objective measures or just subjective measures are being evaluated in terms of a set of alternatives based upon multiple criteria, organized in a hierarchical structure. At the top level, the criteria are evaluated or weighted, and at the bottom level, the alternatives are measured against each criterion. The decision maker assesses their evaluation by making pairwise comparisons in which every pair is subjectively or objectively compared. The subjective method involves a nine-point scale that we present later in Table 4.5.

We only desire to briefly discuss the elements in the framework of AHP. This can be described as a method to decompose a problem into subproblems. In most decisions, the decision maker has a choice among many alternatives. Each alternative has a set of attributes or characteristics that can be measured, either subjectively or objectively. We will call these attributes, criteria.

The attribute elements of the hierarchal process can relate to any aspect of the decision problem – tangible or intangible, carefully measured or roughly estimated, well or poorly understood – anything at all that applies to the decision at hand.

We simply state that in order to perform AHP we need an objective and a set of alternatives, each with criteria (attributes) to compare. Once the hierarchy is built, the decision makers systematically evaluate the various elements pairwise (by comparing them to one another two at a time), with respect to their impact on an element above them in the hierarchy. In making the comparisons, the decision makers can use concrete data about the elements, but they typically use their judgments about the elements' relative meaning and importance. It is the essence of the AHP that human judgments, and not just the underlying information, can be used in performing the evaluations.

The AHP converts these evaluations to numerical values that can be processed and compared over the entire range of the problem. A numerical weight or priority is derived for each element of the hierarchy, allowing diverse and often incommensurable elements to be compared to one another in a rational and consistent way. This capability distinguishes the AHP from other decision-making techniques.

In the final step of the process, numerical priorities are calculated for each of the decision alternatives. These numbers represent the alternatives' relative ability to achieve the decision goal, so they allow a straightforward consideration of the various courses of action (COA).

While it can be used by individuals working on straightforward decisions, the AHP is most useful where teams of people are working on complex problems, especially those with high stakes, involving human perceptions and judgments, whose resolutions have long-term repercussions. It has unique advantages when important elements of the decision are difficult to quantify or compare, or where communication among team members is impeded by their different specializations, terminologies, or perspectives.

Decision situations to which the AHP can be applied include the following where we desire ranking:

- Choice: The selection of one alternative from a given set of alternatives, usually where there are multiple decision criteria involved.

- Ranking: Putting a set of alternatives in order from most to least desirable.

- Prioritization: Determining the relative merit of members of a set of alternatives, as opposed to selecting a single one or merely ranking them.

- Resource allocation: Apportioning resources among a set of alternatives.

- Benchmarking: Comparing the processes in one's own organization with those of other best-of-breed organizations.

- Quality management: Dealing with the multidimensional aspects of quality and quality improvement.

- Conflict resolution: Settling disputes between parties with apparently incompatible goals or positions.

4.4.2 Methodology of the Analytic Hierarchy Process

The procedure for using the AHP can be summarized as:

Step 1. Build the hierarchy for the decision

Goal	Select the best alternative
Criteria	c1, c2, c3, ..., cm
Alternatives	a1, a2, a3, ..., an

Step 2. Judgments and comparison

Build a numerical representation using a nine-point scale in pairwise comparisons for the attribute criteria and the alternatives. The goal, in AHP, is to obtain a set of eigenvectors of the system that measures the importance with respect to the criterion. We can put these values into a matrix or table based on the values from Table 4.5.

We must ensure that this pairwise matrix is consistent according to Saaty's scheme to compute the consistency ratio (*CR*). The value of *CR* must be less than or equal to 0.1 to be considered consistent. Saaty's computed the random index, *RI*, for random matrices for up to ten criteria.

n	1	2	3	4	5	6	7	8	9	10
RI	0	0	0.52	0.89	1.1	1.24	1.35	1.4	1.45	1.49

Next, we approximate the largest eigenvalue, λ, using the power method (Burden and Faires, 2013). We compute the consistency index, *CI*, using the formula:

$$CI = (\lambda - n)/(n-1)$$

TABLE 4.5

Saaty's Nine-Point Scale

Intensity of Importance in Pairwise Comparisons	Definition
1	Equal importance
3	Moderate importance
5	Strong importance
7	Very strong importance
9	Extreme importance
2,4,6,8	For comparing between the above
Reciprocals of above	In comparison of elements i and j, if i is 3 compared to j, then j is 1/3 compared to i.
Rationale	Force consistency; measure values available

Then we compute the *CR* using:

$$CR = CI/RI$$

If $CR \leq 0.1$, then our pairwise comparison matrix is consistent and we may continue the AHP process. If not, we must go back to our pairwise comparison and fix the inconsistencies until the $CR \leq 0.1$. In general, the consistency ensures that if $A > B$, $B > C$, that $A > C$ for all *A*, *B*, and *C* all of which can be criteria or alternatives related by pairwise comparisons.

Step 3. Finding all the eigenvectors combined in order to obtain a comparative ranking.

Step 4. After the $m \times 1$ criterion weights are found and the $n \times m$ matrix for *n* alternatives by *m* criterion, we use matrix multiplication to obtain the $n \times 1$ final rankings.

Step 5. We order the final ranking.

4.4.3 Strengths and Limitations of AHP

Like all modeling methods, the AHP has strengths and limitations.

The main advantage of the AHP is its ability to rank choices in the order of their effectiveness in meeting conflicting objectives. If the judgments made about the relative importance of criteria and those about the alternatives' ability to satisfy those objectives have been made in good faith and effort, then the AHP calculations lead to the logical consequence of those judgments. It is quite hard, but not impossible, to manually change the pairwise judgments to get some predetermined result. A further strength of the AHP is its ability to detect inconsistent judgments in pairwise comparisons using the *CR* value.

The limitations of the AHP are that it only works because the matrices are all of the same mathematical form – known as a positive reciprocal matrix. The reasons for this are explained in Saaty's book, which is not for the mathematically daunted, so we will simply state that point. To create such a matrix requires that, if we use the number 9 to represent "*A* is absolutely more important than *B*", then we have to use 1/9 to define the relative importance of *B* with respect to *A*. Some people regard that as reasonable; others do not.

Some suggest a drawback is in the possible scaling. However, understanding that the final values obtained simply say that one alternative is relatively better than another alternative. For example, if the AHP values for alternatives {*A*, *B*, *C*} found were (0.392, 0.406, 0.204) then they only imply that alternatives *A* and *B* are about equally good at approximately 0.4, while *C* is worse at 0.2. It does not mean that A and B are twice as good as C.

In less clear-cut cases, it would not be a bad thing to change the rating scale and see what difference it makes. If one option consistently scores well with different scales, it is likely to be a very robust choice.

In short, the AHP is a useful technique for discriminating between competing options in the light of a range of objectives to be met. The calculations are not complex, and while the AHP relies on what might be seen as a mathematical trick, you don't need to understand the mathematics to use the technique. Be aware that it only shows relative values.

Although AHP has been used in many applications of the public and private sectors, Hartwich (1999) noted several limitations. First and foremost, AHP was criticized for not providing sufficient guidance about structuring the problem to be solved, forming the levels of the hierarchy for criteria and alternatives, and aggregating group opinions when team members are geographically dispersed or are subject to time constraints. Team members may carry out rating items individually or as a group. As the levels of hierarchy increase, so does the difficulty and time it takes to synthesize weights. One remedy in preventing these problems is by conducting "AHP Walk-throughs" (i.e., meetings of decision-making participants who review the basics of the AHP methodology and work through examples so that concepts are thoroughly and easily understood).

Another critique of AHP is the "rank reversal" problem, i.e., changes in the importance ratings whenever criteria or alternatives are added to or deleted from the initial set of alternatives compared. Several modifications to AHP have been proposed to cope with this and other related issues. Many of the enhancements involved ways of computing, synthesizing pairwise comparisons, and/or normalizing the priority and weighting vectors. We mention now that Technique for Order of Preference by Similarity to Ideal Solution (TOPSIS) corrects this rank reversal issue.

4.4.4 Sensitivity Analysis

Since AHP, at least in the pairwise comparisons, is based upon subjective inputs using the nine-point scale then sensitivity analysis is extremely important. Leonelli (2012), in his master's thesis, outlines procedures for sensitivity analysis to enhance decision support tools including numerical incremental analysis of a weight, probabilistic simulations, and mathematical models. How often do we change our minds about the relative importance of an object, place, or thing? Often enough that we should alter the pairwise comparison values to determine how robust our rankings are in the AHP process. We suggest doing enough sensitivity analysis to find the "break-point" values, if they exist, of the decision maker weights that change the rankings of our alternatives. Since the pairwise comparisons are subjective matrices compiled using the Saaty method, we suggest as a minimum a "trial and error" sensitivity analysis using the numerical incremental analysis of the weights.

Chen and Kocaoglu (2008) grouped sensitivity analysis into three main groups: numerical incremental analysis, probabilistic simulations, and mathematical models. The numerical incremental analysis, also known as one-at-a-time (OAT) or "trial and error", works by incrementally changing one parameter at a time, finding the new solution and showing graphically how the ranks change.

Monte Carlo simulation (Butler et al, 1997) has been used that allows changes to weights and explores the effects on the alternative rankings. Modeling may be used when it is possible to express the relationship between the input data and the solution results.

We used Equation (4.3) (Alinezhad and Amini, 2011) for adjusting weights.

4.4.5 Illustrative Examples with AHP

Example 4.5: Car Selection with AHP

We revisit Car Selection with our raw data presented in Table 4.2 to illustrate AHP in selecting the best alternative based upon pairwise comparisons of the decision criteria.

Step 1. Build the hierarchy and prioritize the criterion from your highest to lowest priority.

Goal	Select the best car
Criteria	c1, c2, c3, …, cm
Alternatives	a1, a2, a3, …, an

For our cars example, we choose the priority as follows: Cost, MPG City, Safety, Reliability, MPG Highway, Performance, and Interior and Style. Putting these in a priority order allows for an easier assessment of the pairwise comparisons. We used an Excel template prepared for these pairwise comparisons.

Step 2. Perform the pairwise comparisons using Saaty's nine-point scale. We use an Excel template created to organize the pairwise comparisons.

	A		B	More Important	Intensity (1-9)
			Element		
1	Cost	compared with	MPG_city	A	2
2			MPG_HW	A	2
3			Safety	A	3
4			Reliability	A	4
5			Performance	A	5
6			Interior & Style	A	6
7					
1	MPG_city	compared with	MPG_HW	A	2
2			Safety	A	3
3			Reliability	A	4
4			Performance	A	5
5			Interior & Style	A	5
6					
1	MPG_HW	comp. with	Safety	A	2
2			Reliability	A	2
3			Performance	A	3
4			Interior & Style	A	3
5					
1	Safety	comp. with	Reliability	A	1
2			Performance	A	2
3			Interior & Style	A	3
4					
1	Reliability vs		Performance	A	2
2			Interior & Style	A	3
3					
1	Performance vs		Interior & Style	A	2
2					

This yields the following decision criterion matrix,

	Cost	G City	MPG HW	Safety	Reliability	Performance	Interior and Style
Cost	1	2	2	3	4	5	6
MPG City	0.5	1	2	3	4	5	5
MPG HW	0.5	0.5	1	2	2	3	3
Safety	0.3333	0.333	0.5	1	1	2	3
Reliability	0.25	0.25	0.5		1	2	3
Performance	0.2	0.2	0.333	0.5	1	1	2
Interior and Style	0.166	0.2	0.333	0.333	0.333	0.5	1

We check the CR to ensure it is less than 0.1. For our pairwise decision matrix, the $CR = 0.00695$. Since the $CR < 0.1$, we continue.

We find the *eigenvector* for the decision weights:

cost	0.342407554
city	0.230887543
hw	0.151297361
safety	0.094091851
reliability	0.080127732
performance	0.055515667
Interior and Style	0.045672293

Step 3. For the alternatives, we either have the data as we obtained it for each car under each decision criterion or we can use pairwise comparisons by criteria for how each car fares versus its competitors. In this example, we take the raw data from before except now we will use 1/*cost* to replace *cost* before we normalize the columns.

We have other options for dealing with a criteria and variable like *cost*. Thus, we have three COA: (1) use 1/*cost* to replace *cost*, (2) use a pairwise comparison using the nine-point scale, or (3) remove *cost* from a criterion and a variable, run the analysis, and then do a *benefit/cost* ratio to re-rank the results.

Step 4. We multiply the matrix of the normalized raw data from Consumer Reports and the matrix of weights to obtain the rankings. Using COA (1) in Step 3, we obtain the following results:

Car	Value
Fusion	0.180528
Camry	0.178434
Prius	0.171964
Sonata	0.168776
Leaf	0.154005
Volt	0.146184

Fusion is our first choice, followed by Camry, Prius, Sonata, Leaf, and Volt.

If we use method COA (2) in Step 3, then within the final matrix, we replace the actual costs with these pairwise results ($CR = 0.031$):

Prius	0.107059
Fusion	0.073259
Volt	0.046756
Camry	0.465277
Sonata	0.256847
Leaf	0.050802

Then we obtain the ranked results as:

Camry	0.270679
Sonata	0.203471
Fusion	0.146001
Prius	0.141518
Leaf	0.12178
Volt	0.116551

If we do COA (3) in Step 3, then this method requires us to redo the pairwise criterion matrix without the cost criteria. These weights are:

City MPG	0.363386
HW MPG	0.241683
Safety	0.159679
Reliability	0.097
Performance	0.081418
Interior/ Style	0.056834

We normalize the original costs from Table 4.2 and divide these ranked values by the normalized *cost* to obtain a *cost/benefit* value. These are shown in ranked order:

Camry	1.211261
Fusion	1.178748
Prius	1.10449
Sonata	1.06931
Leaf	0.821187
Volt	0.759482

Sensitivity Analysis

We alter our decision pairwise values to obtain a new set of decision weights to use in COA (1) from Step 3 to obtain new results: Camry, Fusion, Sonata, Prius, Leaf, and Volt. The new weights and model's results are:

Cost	0.311155922
MPG City	0.133614062
MPG HW	0.095786226
Performance	0.055068606
Interior	0.049997069
Safety	0.129371535
Reliability	0.225006578

Alternatives	Values	
Prius	0.10882648	4
Fusion	0.11927995	2
Volt	0.04816882	5
Camry	0.18399172	1
Sonata	0.11816156	3
Leaf	0.04357927	6

The resulting values have changed but not the relative rankings of the cars. Again, we recommend using sensitivity analysis to find a "break point", if one exists.

We systemically varied the cost weights using Equation (4.1) with increments of (±) 0.05. We potted the results to show the approximate break point of the criteria cost as weight of cost +0.1 as shown in Figure 4.1.

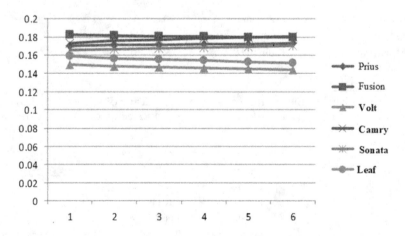

FIGURE 4.1

Camry overtakes Fusion as the top alternative as we change the weight of *Cost*.

Example 4.6: Kite Network Revisited with AHP

Assume all we have are the outputs from ORA: A Toolkit for Dynamic Network Analysis and Visualization, which we do not show here due to the volume of output produced. We take the metrics from ORA and normalize each column. The columns for each criterion are placed in a matrix X with entries, x_{ij}. We define w_j as the weights for each criterion.

Next, we assume we can obtain a pairwise comparison matrix from the decision maker concerning the criterion. We use the output from ORA and normalize the results for AHP to rate the alternatives within each criterion. We provide a sample pairwise comparison matrix for weighting the criterion from the Kite example using Saaty's nine-point scale. The *CR* is 0.0828, which is less than 0.1, so our pairwise matrix is consistent and we continue.

Pairwise Comparison Matrix

	Central	Eigenvector	In-degree	Out-degree	Information Centrality	Betweenness
Central	1	3	2	2	½	1/3
Eigenvector	1/3	1	1/3	1	2	½
In-degree	½	3	1	½	½	¼
Out-degree	½	½	1	1	¼	¼
Information Centrality	2	2	4	4	1	1/3
Betweenness	3	2	4	4	3	1

We obtain the steady state values that will be our criterion weights, where the sum of the weights equals 1.0. There exist many methods to obtain these weights. The methods used here are the power method from numerical analysis (Burden and Faires, 2013) and discrete dynamical systems (Fox, 2012; Giordano et al., 2013).

0.1532	0.1532	0.1532	0.1532	0.1532	0.1532
0.1450	0.1450	0.1450	0.1450	0.1450	0.1450
0.1194	0.1195	0.1194	0.1194	0.1194	0.1194
0.0672	0.0672	0.0672	0.0672	0.0672	0.0672
0.1577	0.1577	0.1577	0.1577	0.1577	0.1577
0.3575	0.3575	0.3575	0.3575	0.3575	0.3575

These values provide the weights for each criterion: *centrality* = 0.1532, *eigenvectors* = 0.1450, *in-centrality* = 0.1194, *out-centrality* = 0.0672, *information centrality* = 0.1577, and *betweenness* = 0.3575.

We multiply the matrix of the weights and the normalized matrix of metrics from ORA to obtain our output and ranking:

Node	AHP Value	Rank
Susan	0.160762473	2
Steven	0.133201647	3
Sarah	0.113388361	4

Tom	0.075107843	6
Claire	0.075107843	6
Fred	0.060386019	8
David	0.060386019	8
Claudia	0.177251415	1
Ben	0.109606727	5
Jennifer	0.034801653	10

For this example with AHP, Claudia, *cl*, is the key node. However, the bias of the decision maker is important in the analysis of the criterion weights. The criterion, "Betweenness", is two to three times more important than the other criterion.

Sensitivity Analysis

Changes in the pairwise decision criterion cause fluctuations in the key nodes. We change our pairwise comparison so that "Betweenness" is not so dominant a criterion.

With these slight pairwise changes, we now find that Susan is ranked first, followed by Steven, and then Claudia. The AHP process is sensitive to changes in the criterion weights. We vary *betweenness* in increments of 0.05 to find the break point.

	Centrality	IN	OUT	Eigen	EigenC	Close	IN-Close	Betw	INFO Cen.
t	0.111111	0.111111	0.111111	0.114399	0.114507	0.100734	0.099804	0.019408	0.110889
c	0.111111	0.111111	0.111111	0.114399	0.114507	0.100734	0.099804	0.019408	0.108891
f	0.083333	0.083333	0.083333	0.093758	0.094004	0.097348	0.09645	0	0.097902
s	0.125	0.138889	0.111111	0.137528	0.137331	0.100734	0.111826	0.104188	0.112887
su	0.180556	0.166667	0.194444	0.175081	0.174855	0.122743	0.107632	0.202247	0.132867
st	0.138889	0.138889	0.138889	0.137528	0.137331	0.112867	0.111826	0.15526	0.123876
d	0.083333	0.083333	0.083333	0.093758	0.094004	0.097348	0.107632	0	0.100899
cl	0.083333	0.083333	0.083333	0.104203	0.104062	0.108634	0.107632	0.317671	0.110889
b	0.055556	0.055556	0.055556	0.024123	0.023985	0.088318	0.087503	0.181818	0.061938
j	0.027778	0.027778	0.027778	0.005223	0.005416	0.070542	0.069891	0	0.038961

10 alternatives and 9 attributes or criterion
Criterion weights

w1	0.034486
w2	0.037178
w3	0.045778
w4	0.398079
w5	0.055033
w6	0.086323
w7	0.135133
w8	0.207991

With these slight pairwise changes, we now find Susan is now ranked first, followed by Steven and then Claudia. The AHP process is sensitive to changes in the criterion weights. We vary *betweenness* in increments of 0.05 to find the break point.

Tom	0.098628	Susan	0.161609
Claire	0.098212	Steven	0.133528
Fred	0.081731	Claudia	0.133428
Sarah	0.12264	Sarah	0.12264
Susan	0.161609	Tom	0.098628
Steven	0.133528	Claire	0.098212
David	0.083319	David	0.083319
Claudia	0.133428	Fred	0.081731
Ben	0.0645	Ben	0.0645
Jennifer	0.022405	Jennifer	0.022405

Further, sensitivity analysis of the nodes is provided in Figure 4.2.

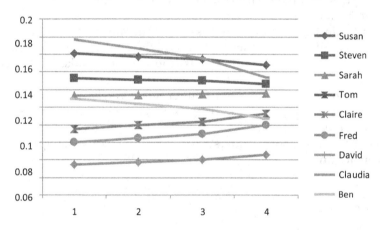

FIGURE 4.2
Sensitivity analysis for Nodes varying only *Betweenness*.

We varied the weight of the criterion *Betweenness* by lowering it by 0.05 each iteration and increasing the other weights using Equation (4.1). We see the Claudia and Susan change as the top node when we reduce *Betweenness* by 0.1.

4.5 Technique of Order Preference by Similarity to Ideal Solution (TOPSIS)

4.5.1 Description and Uses

The TOPSIS is a multi-criteria decision analysis method, which was originally developed in a dissertation from Kansas State University

(Hwang and Yoon, 1981). It has been further developed by others (Yoon, 1987; Hwang et al, 1993). TOPSIS is based on the concept that the chosen alternative should have the shortest geometric distance from the positive ideal solution and the longest geometric distance from the negative ideal solution. It is a method of compensatory aggregation that compares a set of alternatives by identifying weights for each criterion, normalizing the scores for each criterion, and calculating the geometric distance between each alternative and the ideal alternative, which is the best score in each criterion. An assumption of TOPSIS is that the criteria are monotonically increasing or decreasing. Normalization is usually required as the parameters or criteria are often of incompatible dimensions in multi-criteria problems. Compensatory methods such as TOPSIS allow trade-offs between criteria, where a poor result in one criterion can be negated by a good result in another criterion. This provides a more realistic form of modeling than non-compensatory methods, which include or exclude alternative solutions based on hard cutoffs.

We only desire to briefly discuss the elements in the framework of TOPSIS. TOPSIS can be described as a method to decompose a problem into subproblems. In most decisions, the decision maker has a choice among many alternatives. Each alternative has a set of attributes or characteristics that can be measured, either subjectively or objectively. The attribute elements of the hierarchal process can relate to any aspect of the decision problem – tangible or intangible, carefully measured or roughly estimated, well or poorly understood – anything at all that applies to the decision at hand.

4.5.2 Methodology

The TOPSIS process is carried out as follows:

Step 1. Create an evaluation matrix consisting of m alternatives and n criteria, with the intersection of each alternative and criterion given as x_{ij} giving us a matrix $(X_{ij})_{m \times n}$.

$$D = \begin{array}{c} \\ A_1 \\ A_2 \\ A_3 \\ \cdot \\ \cdot \\ \cdot \\ A_m \end{array} \begin{bmatrix} x_1 & x_2 & x_3 & \cdots & x_n \\ x_{11} & x_{12} & x_{13} & \cdots & x_{1n} \\ x_{21} & x_{22} & x_{23} & \cdots & x_{2n} \\ x_{31} & x_{32} & x_{33} & \cdots & x_{3n} \\ \cdot & \cdot & \cdot & & \cdot \\ \cdot & \cdot & \cdot & & \cdot \\ \cdot & \cdot & \cdot & & \cdot \\ x_{m1} & x_{m2} & x_{m3} & \cdots & x_{mn} \end{bmatrix}$$

Step 2. The matrix shown as D above then is normalized to form the matrix $R = (R_{ij})_{m \times n}$ as shown using the normalization method

$$r_{ij} = \frac{x_{ij}}{\sqrt{\sum x_{ij}^2}}$$

for $i = 1,2\ldots, m; j = 1,2, \ldots n$

Step 3. Calculate the weighted normalized decision matrix. First, we need the weights. Weights can come from either the decision maker or by computation.

Step 3a. Use either the decision maker's weights for the attributes x_1, $x_2, \ldots, x_n$ or compute the weights through the use of Saaty's (1980) AHP decision maker weights method to obtain the weights as the eigenvector to the attributes versus attribute pairwise comparison matrix.

$$\sum_{j=1}^{n} w_j = 1$$

The sum of the weights over all attributes must equal 1 regardless of the method used.

Step 3b. Multiply the weights to each of the column entries in the matrix from *Step 2* to obtain the matrix, T.

$$T = (t_{ij})_{m \times n} = (w_j r_{ij})_{m \times n}, i = 1, 2, \ldots, m$$

Step 4. Determine the worst alternative (A_w) and the best alternative (A_b): examine each attribute's column and select the largest and smallest values appropriately. If the values imply larger is better (profit), then the best alternatives are the largest values, and if the values imply smaller is better (such as cost), then the best alternative is the smallest value.

$$A_w = \left\{ \langle \max(t_{ij} | i = 1,2,\ldots,m \,|\, j \in J_-), \langle \min(t_{ij} | i = 1, 2, \ldots, m) \,|\, j \in J_+ \rangle \right\}$$

$$\equiv \left\{ t_{wj} | j = 1, 2, \ldots, n \right\},$$

$$A_{wb} = \left\{ \langle \min(t_{ij} | i = 1,2,\ldots,m \,|\, j \in J_-), \langle \max(t_{ij} | i = 1, 2, \ldots, m) \,|\, j \in J_+ \rangle \right\}$$

$$\equiv \left\{ t_{bj} | j = 1, 2, \ldots, n \right\},$$

where:

$J_+ = \{j = 1,2,\ldots n | j)$ associated with the criteria having a positive impact

$J_- = \{j = 1,2,\ldots n | j)$ associated with the criteria having a negative impact

We suggest that if possible, make all entry values in terms of positive impacts.

Step 5. Calculate the L2-distance between the target alternative i and the worst condition A_w

$$d_{iw} = \sqrt{\sum_{j=1}^{n}(t_{ij} - t_{wj})^2}, i = 1, 2, \ldots, m$$

and then calculate the distance between the alternative i and the best condition A_b

$$d_{ib} = \sqrt{\sum_{j=1}^{n}(t_{ij} - t_{bj})^2}, i = 1, 2, \ldots m$$

where d_{iw} and d_{ib} are L2-norm distances from the target alternative i to the worst and best conditions, respectively.

Step 6. Calculate the similarity to the worst condition:

$$s_{iw} = \frac{d_{iw}}{(d_{iw} + d_{ib})}, 0 \leq s_{iw} \leq 1, i = 1, 2, \ldots, m$$

$S_{iw} = 1$ if and only if the alternative solution has the worst condition

$S_{iw} = 0$ if and only if the alternative solution has the best condition

Step 7. Rank the alternatives according to their value from S_{iw} ($i = 1, 2, \ldots, m$).

4.5.3 Normalization

Two methods of normalization that have been used to deal with incongruous criteria dimensions are linear normalization and vector normalization.

Normalization can be calculated as in *Step 2* of the TOPSIS process above. Vector normalization was incorporated with the original development of the TOPSIS method (Yoon, 1987) and is calculated using the following formula:

$$r_{ij} = \frac{x_{ij}}{\sqrt{\sum x_{ij}^2}} \text{ for } i = 1, 2 \ldots, m; j = 1, 2, \ldots n$$

In using vector normalization, the nonlinear distances between single-dimension scores and ratios should produce smoother trade-offs (Hwang and Yoon, 1981).

Let's suggest two options for the weights in Step 3. First, the decision maker might actually have a weighting scheme that they want the analyst to use. If not, we suggest using Saaty's nine-point pairwise method developed for the AHP (Saaty, 1980). We refer the reader to our discussion in the AHP section for the decision weights using the Saaty's nine-point scale and pairwise comparisons. In TOPSIS, we have the following scheme.

Objective Statement: This is the decision desired:

Alternatives: 1, 2, 3, ..., *n*

For each of the alternatives, there are criteria (attributes) to compare: Criteria (or Attributes): *c1, c2,..., cm*

Once the hierarchy is built, the decision maker(s) systematically evaluate its various elements pairwise (by comparing them to one another two at a time), with respect to their impact on an element above them in the hierarchy. In making the comparisons, the decision makers can use concrete data about the elements, but they typically use their judgments about the elements' relative meaning and importance. It is the essence of the TOPSIS that human judgments, and not just the underlying information, can be used in performing the evaluations.

TOPSIS converts these evaluations to numerical values that can be processed and compared over the entire range of the problem. A numerical weight or priority is derived for each element of the hierarchy, allowing diverse and often incommensurable elements to be compared to one another in a rational and consistent way. This capability distinguishes the TOPSIS from other decision-making techniques.

In the final step of the process, numerical priorities or ranking is calculated for each of the decision alternatives. These numbers represent the alternatives' relative ability to achieve the decision goal, so that they allow a straightforward consideration of the various courses of action.

While it can be used by individuals working on straightforward decisions, TOPSIS is most useful where teams of people are working on complex problems, especially those with high stakes, involving human perceptions and judgments, whose resolutions have long-term repercussions. It has unique advantages when important elements of the decision are difficult to quantify or compare, or where communication among team members is impeded by their different specializations, terminologies, or perspectives.

Decision situations to which the TOPSIS might be applied are identical to what we presented earlier for AHP:

- Choice: The selection of one alternative from a given set of alternatives, usually where there are multiple decision criteria involved.
- Ranking: Putting a set of alternatives in order from most to least desirable.

- Prioritization: Determining the relative merit of members of a set of alternatives, as opposed to selecting a single one or merely ranking them.
- Resource allocation: Apportioning resources among a set of alternatives.
- Benchmarking: Comparing the processes in one's own organization with those of other best-of-breed organizations.
- Quality management: Dealing with the multidimensional aspects of quality and quality improvement.
- Conflict resolution: Settling disputes between parties with apparently incompatible goals or positions.

4.5.4 Strengths and Limitations

TOPSIS is based on the concept that the chosen alternative should have the shortest geometric distance from the positive ideal solution and the longest geometric distance from the negative ideal solution. It is a method of compensatory aggregation that compares a set of alternatives by identifying weights for each criterion, normalizing scores for each criterion, and calculating the geometric distance between each alternative and the ideal alternative, which is the best score in each criterion. An assumption of TOPSIS is that the criteria are monotonically increasing or decreasing. Normalization is usually required as the parameters or criteria are often of incongruous dimensions in multi-criteria problems. Compensatory methods such as TOPSIS allow trade-offs between criteria, where a poor result in one criterion can be negated by a good result in another criterion. This provides a more realistic form of modeling than non-compensatory methods, which include or exclude alternative solutions based on hard cutoffs. TOPSIS corrects the rank reversal that was a limitation in strictly using the AHP method.

TOPSIS also allows the user to state which of the criteria are maximized and which are minimized for better results. In the late 1980s, TOPSIS was a department of defense standard for performing the selection of systems across all branches in tight budget years.

4.5.5 Sensitivity Analysis

The decision weights are subject to sensitivity analysis to determine how they affect the final ranking. The same procedures discussed in Section 4.4 are valid here. Sensitivity analysis is essential to good analysis. Additionally, Alinezhad et al. (2011) suggests sensitivity analysis for TOPSIS for changing an attribute weight. We will again use Equation (4.3) in our sensitivity analysis.

4.5.6 Illustrate Examples with TOPSIS

Example 4.7: Car Selection with TOPSIS (see Table 4.2)

We might assume that our decision maker weights from the AHP section are still valid for our use.

Weights from before:

Cost	0.38960838
MPG City	0.11759671
MPGHW	0.04836533
Performance	0.0698967
Interior	0.05785692
Safety	0.10540328
Reliability	0.21127268

	cost	MPG_city	MPG_HW	Perf.	Interior	safety	reliability	N/A
Cost	1	4	6	5	6	4	2	0
MPG_city	0.25	1	6	3	5	1	0.333333333	0
MPG_HW	0.166667	0.166667	1	0.5	0.5	0.333333	0.25	0
Perf.	0.2	0.333333	2	1	2	0.5	0.333333333	0
Interior	0.166667	0.2	2	0.5	1	0.5	0.333333333	0
safety	0.25	1	3	2	2	1	0.5	0
reliability	0.5	3	4	3	3	2	1	0
N/A	0	0	0	0	0	0	0	1

We use the identical data from the car example from AHP, but we apply Steps 3–7 from TOPSIS to our data. We are able to keep the cost data and just inform TOPSIS that a smaller cost is better. We obtained the following rank ordering of the cars: Camry, Fusion, Prius, Sonata, Volt, and Leaf.

4	0.82154128	Camry
2	0.74622988	Fusion
1	0.72890117	Prius
5	0.70182382	Sonata
6	0.15580913	Leaf
3	0.11771999	Volt

It is critical to perform sensitivity analysis on the weights to see how they affect the final ranking. This time we work toward finding the break point where the order of cars actually changes. Since cost is the largest criterion weight, we vary it using Equation (4.3) in increments of 0.05. We see from Figure 4.3 that the Fusion overtakes Camry when cost is decreased by about 0.1, which allows reliability to overtake cost as the dominate-weighted decision criterion.

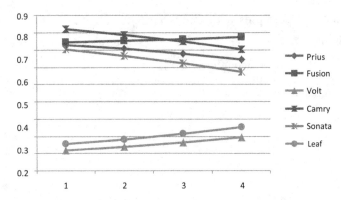

FIGURE 4.3
TOPSIS values of the cars by varying the weight for cost incrementally by −0.05
each of four increments along the *X-axis*.

Example 4.8: Social Networks with TOPSIS

We revisit the Kite Network with TOPSIS to find influences in the network. We present the output from ORA that we used in Table 4.6.

TABLE 4.6

Summary of ORA's Output for Kite Network

	IN	OUT	Eigen	EigenL	Close	IN-Close	Betweenness	INF Center
Tom	0.4	0.4	0.46	0.296	0.357	0.357	0.019	0.111
Claire	0.4	0.4	0.46	0.296	0.357	0.357	0.019	0.109
Fred	0.3	0.3	0.377	0.243	0.345	0.345	0	0.098
Sarah	0.5	0.4	0.553	0.355	0.357	0.4	0.102	0.113
Susan	0.6	0.7	0.704	0.452	0.435	0.385	0.198	0.133
Steven	0.5	0.5	0.553	0.355	0.4	0.4	0.152	0.124
David	0.3	0.3	0.377	0.243	0.345	0.385	0	0.101
Claudia	0.3	0.3	0.419	0.269	0.385	0.385	0.311	0.111
Ben	0.2	0.2	0.097	0.062	0.313	0.313	0.178	0.062
Jennifer	0.1	0.1	0.021	0.014	0.25	0.25	0	0.039

We use the decision weights from AHP (unless a decision maker gives
us their own weights) and find the eigenvectors for our eight metrics as:

w1	0.034486
w2	0.037178
w3	0.045778
w4	0.398079
w5	0.055033
w6	0.086323
w7	0.135133
w8	0.207991

We take the metrics from ORA and perform Steps 2–7 of TOPSIS to obtain the results:

S+	S-	C	
0.0273861	0.181270536	0.86875041	SUSAN
0.0497878	0.148965362	0.749499497	STEVEN
0.0565358	0.14154449	0.714581437	SARAH
0.0801011	0.134445151	0.626648721	TOM
0.0803318	0.133785196	0.624822765	CLAIRE
0.10599	0.138108941	0.565790826	CLAUDIA
0.1112243	0.12987004	0.538668909	DAVID
0.1115873	0.128942016	0.536076177	FRED
0.1714404	0.113580988	0.398499927	BEN
0.2042871	0.130399883	0.389617444	JENNIFER

We rank order the final output from TOPSIS as shown in the last column above. We interpret the results as follows: the key node is *Susan* followed by *Steven, Sarah, Tom,* and *Claire.*

Sensitivity Analysis

We used Equation (4.3) and systemically altered the value of the largest criteria weight, *EigenL*, and depicted this in Figure 4.4.

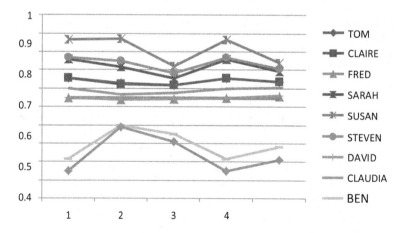

FIGURE 4.4
Sensitivity analysis plot as a function of varying *EigenL* weight in increments of −0.05 units. We note that Susan remains the most influential node.

Comparison of Results for the Kite Network

We have also used the two other MADM methods to rank order our nodes in previous work in SNA (Fox and Everton, 2013). When we applied data envelopment analysis and AHP to compare to TOPSIS, we obtained the results displayed in Table 4.6 for the Kite Network (Table 4.7).

TABLE 4.7

MADM Applied to Kite Network

Node	SAW	TOPSIS Value (rank)	DEA Efficiency Value (rank)	AHP Value (rank)
Susan	0.046 (1)	0.862 (1)	1 (1)	0.159 (2)
Sarah	0.021 (4)	0.675 (3)	0.786 (2)	0.113 (4)
Steven	0.026 (3)	0.721 (2)	0.786 (2)	0.133 (3)
Claire	0.0115 (7)	0.649 (4)	0.653 (4)	0.076 (6)
Fred	0.0115 (7)	0.446 (8)	0.653 (4)	0.061 (8)
David	0.031 (2)	0.449 (7)	0.536 (8)	0.061 (8)
Claudia	0.012 (8)	0.540 (6)	0.595 (6)	0.176 (1)
Ben	0.018 (5)	0.246 (9)	0.138 (9)	0.109 (5)
Jennifer	0.005 (10)	0 (10)	0.030 (10)	0.036 (10)
Tom	0.0143 (6)	0.542 (5)	0.553 (7)	0.076 (6)

It might be useful to use this table as input for another round of one of these presented methods and then use sensitivity analysis.

Chapter 4 Exercises

In each problem, use SAW and then TOPSIS to find the ranking under these weighted conditions:

a. All weights are equal.

b. Choose and state your weights.

4.1 For a given hospital, rank order the procedure using the data below.

	Procedure			
	1	2	3	4
Profit	$200	$150	$100	$80
X-Ray Times	6	5	4	3
Laboratory Time	5	4	3	2

4.2 For a given hospital, rank order the procedure using the data below.

	Procedure			
	1	2	3	4
Profit	$190	$150	$110	980
X-Ray Times	6	5	5	3
Laboratory Time	5	4	3	3

4.3 Rank order the following threats:

Threat Alternatives\ Criterion	Reliability of Threat Assessment	Approximate Associated Deaths (000)	Cost to Fix Damages in Millions	Location Density in Millions	Destructive Psychological Influence	Number of Intelligence-Related Tips
Dirty Bomb Threat	0.40	10	150	4.5	9	3
Anthrax-Bio Terror Threat	0.45	0.8	10	3.2	7.5	12
DC-Road and Bridge Network Threat	0.35	0.005	300	.85	6	8
NY Subway Threat	0.73	12	200	6.3	7	5
DC Metro Threat	0.69	11	200	2.5	7	5
Major Bank Robbery	0.81	0.0002	10	0.57	2	16
FAA Threat	0.70	0.001	5	0.15	4.5	15

4.4 Consider a scenario where we want to move, rank the cities.

City	Affordability of Housing (Average Home Cost in Hundreds of Thousands)	Cultural Opportunities – Events per Month	Crime rate – Number of Reported # Crimes per Month (in Hundreds)	Quality of Schools on Average (Quality Rating Between [0,1])
1	250	5	10	0.75
2	325	4	12	0.6
3	676	6	9	0.81
4	1,020	10	6	0.8
5	275	3	11	0.35
6	290	4	13	0.41
7	425	6	12	0.62
8	500	7	10	0.73
9	300	8	9	0.79

4.5 Consider rating departments at a college.
     The following table is provided:

DMU Departments	Inputs # Faculty	Outputs Student Credit Hours	Outputs Number of Students	Outputs Total Degrees
Unit 1	25	18,341	9,086	63
Unit 2	15	8,190	4,049	23
Unit 3	10	2,857	1,255	31
Unit 4	33	22,277	6,102	31
Unit 5	12	6,830	2,910	19

Formulate and solve the DEA model and rank order the five departments.

4.6 Consider ranking companies within a larger organization. For simplification reasons, we will consider only six companies.

Companies	Inputs # Size of Unit	Output #1	Output #2	Output #3
Unit1	120	18,341	9,086	63
Unit 2	110	8,190	4,049	23
Unit 3	100	2,857	1,255	31
Unit 4	135	22,277	6,102	31
Unit 5	120	6,830	2,910	19
Unit 6	95	5,050	1,835	12

4.7 Given the input-output table below for three hospitals where inputs are number of beds and labor hours in thousands per month and outputs, all measured in hundreds, are patient-days for patients under 14, patient-days for patients between 14 and 65, and patient-days for patients over 65. Determine the efficiency of the three hospitals.

Hospital	Inputs		Outputs		
	1	2	1	2	3
1	5	14	9	4	16
2	8	15	5	7	10
3	7	12	4	9	13

4.8 Resolve Problem 3 with the following inputs and outputs.

Hospital	Inputs		Outputs		
	1	2	1	2	3
1	4	16	6	5	15
2	9	13	10	6	9
3	5	11	5	10	12

4.9 Consider ranking 4 bank branches in a particular city. The inputs are:

Input 1 = labor hours in hundred per month

Input 2 = space used for tellers in hundreds of square feet

Input 3 = supplies used in dollars per month

Output 1 = loan applications per month

Output 2 = deposits made in thousands of dollars per month

Output 3 = checks processed thousands of dollars per month

The following data table is for the bank branches.

Branches	Input 1	Input 2	Input 3	Output 1	Output 2	Output 3
1	15	20	50	200	15	35
2	14	23	51	220	18	45
3	16	19	51	210	17	20
4	13	18	49	199	21	35

4.10 What "best practices" might you suggest to the branches that are less efficient in Problem 9?

Key Terms and Definitions

Analytical hierarchy process: AHP is a technique created by Saaty using a nine-point scale to rank alternatives in a decision process and is useful to get decision maker weights for use in TOPSIS.

Decision matrix: This is the $m \times n$ matrix of the m alternatives by n attributes.

Decision weights (eigenvectors): These are the subjective decision weights that are either provided by the decision maker or computed from the pairwise comparison matrix as eigenvectors to the maximum eigenvalue.

Ideal solution: Although assumed unachievable the ideal and negative ideal solutions are used to compute the ratios of distances from the ideal and negative ideal solution.

Normalization process: The normalization process for TOPSIS differs from other processes in that TOPSIS considers distances.

S^-: This represents the distance of the computed value to the negative ideal solution.

S^+: This represents the distance of the computed value to the ideal solution.

TOPSIS: A technique for order preference by similarity to ideal solution.

References

Alinezhad, A. & A. Amini. (2011). Sensitivity Analysis of TOPSIS technique: The results of change in the weight of one attribute on the final ranking of alternatives. *Journal of Optimization in Industrial Engineering*, 7 (2011), pp. 23–28.

Chen, H. & D. Kocaoglu (2008). A sensitivity analysis algorithm for hierarchical decision models. *European Journal of Operations Research*, 185 (1), pp. 266–288.

Fishburn, P.C. (1967). Additive utilities with incomplete product set: Applications to priorities and assignments. *Operations Research Society of America (ORSA)*. 15, pp. 537–542.

Fox, W. & S. Everton. (2013). Mathematical modeling in social network analysis: Using TOPSIS to find node influences in a social network. *Journal of Mathematics and System Science*, 3 (10), pp. 531–541.

Fox, W.P. (2012). Mathematical modeling of the analytical hierarchy process using discrete dynamical systems in decision analysis. *Computers in Education Journal*, July–Sept, pp. 27–34.

Giordano, F.R., Fox, W. &. Horton S. (2013). *A First Course in Mathematical Modeling* (5th ed.). Boston, MA: Brooks-Cole Publishers.

Hartwich, F. (1999). Weighting of agricultural research results: Strength and limitations of the analytic hierarchy process (AHP). Universitat Hohenheim. Retrieved from https://entwicklungspolitik.uni-hohenheim.de/uploads/media/DP_09_1999_Hartwich_02.pdf

Hwang, C.L. & K. Yoon. (1981). *Multiple Attribute Decision Making: Methods and Applications*. New York: Springer-Verlag.

Hwang, C.L., Y. Lai, & T.Y. Liu. (1993). A new approach for multiple objective decision making. *Computers and Operational Research*, 20, pp. 889–899.

Leonelli, R. (2012). Enhancing a decision support tool with sensitivity analysis. Thesis, University of Manchester, Manchester, UK.

Miller, D.W.; M.K. Starr (1969). *Executive Decisions and Operations Research*. Englewood Cliffs, NJ, U.S.A.: Prentice-Hall, Inc.

Saaty, T. (1980). *The Analytical Hierarchy Process*. United States: McGraw Hill.

Shannon, C. E. and Weaver W. 1948. *A Mathematical Theory of Communication. The Bell System Technical Journal*, 27, 623–656.

Temesi, J. (2006). Int. J. Management and Decision Making, Vol. 7, Nos. 2/3.

Yoon, K. (1987). A reconciliation among discrete compromise situations. *Journal of Operational Research Society*, 38, pp. 277–286.

Zeleny, M., & Cochrane, J. L. (1982). *Multiple criteria decision making McGraw-Hill New York*, 34, 1011–1022.

Additional Readings

Baker T. & Z. Zabinsky. (2011). A multicriteria decision making model for reverse logistics using Analytical Hierarchy Process. *Omega*, (39), pp. 558–573.

Burden, R. & D. Faires. (2013). *Numerical Analysis* (9th ed.). Boston, MA: Cengage Publishers.

Butler, J., J. Jia, & J. Dyer. (1997). Simulation techniques for the sensitivity analysis of multi-criteria decision models. *European Journal of Operations Research*, 103, pp. 531–546.

Callen, J. (1991). Data envelopment analysis: practical survey and managerial accounting applications. *Journal of Management Accounting Research*, 3 (1991), pp. 35–57.

Carley, K.M. (2011). *Organizational Risk Analyzer (ORA)*. Pittsburgh, PA: Center for Computational Analysis of Social and Organizational Systems (CASOS): Carnegie Mellon University.

Charnes, A., W. Cooper, & E. Rhodes. (1978). Measuring the efficiency of decision making units. *European Journal of Operations Research*, 2 (1978), pp. 429–444.

Consumer's Reports Car Guide. (2012). The Editors of Consumer Reports. Cooper, W., L. Seiford, & K. Tone. (2000). *Data Envelopment Analysis*. London, UK: Kluwer Academic Press.

Cooper, W., S. Li, L. Seiford, R.M. Thrall, & J. Zhu. (2001). Sensitivity and stability analysis in DEA: Some recent developments. *Journal of Productivity Analysis*, 15 (3), pp. 217–246.

Fox, W. & S. Everton. (2014). Mathematical Modeling in social network analysis: Using data envelopment analysis and analytical hierarchy process to find node influences in a social network. *Journal of Defense Modeling and Simulation*, 2 (2014), pp. 1–9.

Hurly, W.J. (2001). The analytical hierarchy process: A note on an approach to sensitivity which preserves rank order. *Computers and Operations Research*, 28, pp. 185–188.

Krackhardt, D. (1990). Assessing the political landscape: Structure, cognition, and power in organizations. *Administrative Science Quarterly*, 35, pp. 342–369.

Neralic, L. (1998). Sensitivity analysis in models of data envelopment analysis. *Mathematical Communications*, 3, pp. 41–59.

Thanassoulis, E. (2011). *Introduction to the Theory and Application of Data Envelopment Analysis: A Foundation Text with Integrated Software*. London, UK: Kluwer Academic Press.

Trick, M.A. (1996). Multiple Criteria Decision Making for Consultants. http://mat.gsia.cmu.edu/classes/mstc/multiple/multiple.html. Accessed April 2014.

Trick, M.A. (2014). Data Envelopment Analysis, Chapter 12. http://mat.gsia.cmu.edu/classes/QUANT/NOTES/chap12.pdf. Accessed April 2014.

Winston, W. (1995). *Introduction to Mathematical Programming*. Belmont, CA: Duxbury Press. pp. 322–325.

Zhenhua, G. (2009). The application of DEA/AHP method to supplier selection. *2009 International Conference on Information Management, Innovation Management and Industrial Engineering*, Xi'an, China, pp. 449–451. https://doi.org/10.1109/ICIII.2009.266.

5

Game Theory: Total Conflict

Perhaps you have played or seen a game called rock, paper, scissors between two players. Assume that there is an extra piece of pie available and both want it so the two players decide to play the game to see who wins the piece of pie. In the game, each player might choose rock (a fist), paper (a palm), or scissors (two extended fingers). In the game, rock beats scissors, scissors beat paper, and paper beats rock. The game is a simultaneous game as the player displays their choice, perhaps on the count of three. So, what do you do in order to win? We will revisit this game later in this chapter.

Game theory is a decision-making process where there are at least two players in conflict. Although each player may be in situations that involve both conflict and cooperation, we will first consider the conflict. Let's define a game in which the following assumptions are assumed to be true.

 a. There are at least two players. These players may be individuals, groups, companies, countries, or even specifies.
 b. Each player has a number of strategies, courses of action that might be chosen.
 c. These strategies determine the outcome of the game.
 d. Associated with these strategies are numerical payoffs to each player. These payoffs represent the value of the game's outcome to the different players.
 e. Games are played many times not just once in solving for optimal strategies to play.
 f. Games are simultaneous.

Game theory involves how these players should <u>rationally</u> play the game. Each player plays the game in order to obtain the largest outcome (payoff) possible. Players have control over the payoffs by their choice of strategy to play. The outcome is not just determined by their choice but also by their opponent's choice. Here are a few assumptions concerning the application of game theory:

Assume rational choice: Each player chooses a strategy that enables that player to do the best that he can do, given the opponent knows the strategy that he will follow.

Rational does not imply moral or ethical just that the person makes a reasoned choice to help them win.

          DOI: 10.1201/9781032726885-5

Over the many years teaching game theory, I have had many students who say that trying to do the best is not always how people act. They use a divorce example, where the couple really don't like each other and prefer to harm the opponent rather than get the best outcome for themselves. In those cases, game theory won't apply. But if the player assumes that he/she wants the best result regardless of what their opponent does, then we may still use game theory in the selection of our optimal strategies.

5.1 Introduction to Total Conflict Games

A total conflict game is also known as a zero-sum or constant-sum game. A simultaneous game matrix is a matrix that provides row and column values for two players based upon their chosen strategies. The matrix provides the amount won (+) and amount lost (−) for the row and/or column player. If the sum of the row and column entries is 0, then the game is a zero-sum game. If the sum of the row and column entries is the same constant value for all rows and columns, then the game is a constant sum game and treated as if it were a zero-sum game. An example of this type of game is rock-paper-scissors where the players decide to play rock, paper, or scissors. In this game, rock beats scissors, paper beats rock, and scissors beat paper. We illustrate this game in more detail later in the chapter.

In our textbook, we will use the convention that for the general game the row player is called Rose and the column player is called Colin. We do this for simplicity.

Let's define the total conflict game with the following payoff matrix that has strategy components for both Rose and Colin where Rose has m strategies and Colin has n strategies:

$$
(M,N) = \begin{bmatrix}
(M_{1,1}, N_{1,1}) & (M_{1,2}, N_{1,2}) & \cdots & (M_{1,n}, N_{1,n}) \\
(M_{2,1}, N_{2,1}) & (M_{2,2}, N_{2,2}) & \cdots & (M_{2,n}, N_{2,n}) \\
\cdot & \cdot & \cdots & \cdot \\
\cdot & \cdot & \cdots & \cdot \\
\cdot & \cdot & \cdots & \cdot \\
(M_{m,1}, N_{m,1}) & (M_{m,2}, N_{m,2}) & \cdots & (M_{m,n}, N_{m,n})
\end{bmatrix}
$$

In the special case of zero-sum games, each pair sums to 0. For example, one such pair is $M_{11} + N_{11} = 0$. In the special case of the constant sum game, all pairs sum to the same constant, C. For example, the sum of all $M_{ij} + N_{ij} = C$.

For example, the following is a zero-sum as each pair of entries sums to 0.

		Colin	
		C1	C2
Rose	R1	(2,–2)	(1,–1)
	R2	(3,–3)	(4,–4)

A constant-sum game where all entries sum to 100 could look like the following payoff matrix,

		C1	C2
Rose	R1	(20,80)	(50,50)
	R2	(30, 70)	(60,40)

Next, we begin our discussion of solution methods to such games. We begin with a concept called the Nash equilibrium (see Nash, 1950).

Nash equilibrium is a game theory concept that determines the optimal solution in a noncooperative game in which each player lacks any incentive to change his/her initial strategy. Under the Nash equilibrium, a player does not gain anything from deviating from their initially chosen strategy, assuming the other players also keep their strategies unchanged. A game may include one or multiple Nash equilibria.

Minimax theorem for the two-person zero-sum game guarantees that there is a unique game value and an optimal strategy for each player, so that either player can realize *at least* this value by playing this strategy which may be pure or mixed.

5.2 Models with Pure Strategy Solutions

Given a game matrix, a pure strategy means that you are always making the same choice of strategies for each player. Methods to examine a payoff matrix for pure strategy solution include movement diagram and the saddle point method.

5.2.1 Movement Arrows with Two-Players and a Payoff Matrix

For the row players, values draw an arrow in each column from the smaller value to the larger value. For the column players, values draw an arrow from the smaller value to the larger value on each row.

Example 5.1: Consider the Following Two-Person Game

		Colin	
		C1	C2
Rose	R1	(2,−2)	(1,−1)
	R2	(3,−3)	(4,−4)

R1 is smaller than R2 in C1 so the arrows move from R1 to R2. R1 is smaller than R2 in C2 so the arrow moves from R1 to R2. C1 is smaller than C2 in R1 so the arrows move from C1 to C2. In R2, C2 is smaller than C1 so the arrows move from C2 to C1 in R2.

		Colin	
		C1	C2
Rose	R1	(2,−2)	
			(1,−1)
	R2	(3,−3)	(4,−4)

All arrows point in to (3, −3) at R2, C1. Therefore, we have a pure strategy solution at R2C1 with value (3, −3).

If you only use the row player's information then the movement arrows algorithm becomes:

MOVEMENT ARROWS

In each row, draw an arrow from the larger payoff to the smaller payoff. In each column draw, an arrow from the smaller payoff to the larger payoff.

Example 5.2: Zero-Sum Two-Person Game

		Colin	
		C1	C2
Rose	R1	2	1
	R2	3	4

Solution is 3 for Rose when Rose plays R1 and Colin plays C1. We know that Colin gets −3 because the game is zero-sum.

5.2.2 Saddle Point Method

The saddle point method uses the minimax theorem.

MINIMAX

Take the minimum in each row, Then, take the largest of the minimums. Row Min (take max); Take the maximum in each column and then take the smallest of the maximums. Column max (take min) if they match we have an equilibrium (solution).

Example 5.3: Saddle Point Two-Person Three Strategy each Game

		C1	C2	C3	Row Min	Max of Row Mins
Rose	R1	4	4	10	4	
	R2	2	3	1	1	
	R3	6	5	7	5	5
Column Max		6	5	10		
Min of Col. Max			5			

(Colin across C1 C2 C3)

We find that the maximum of the row minimums is equal to the minimum of the column maximums and is, therefore, a saddle point. The saddle point indicates a pure strategy solution. Here Rose plays.

Example 5.4: Two-Person Two-Strategy Game

		C1	C2	Row Min
Rose	R1	2	2	2 Max
	R2	1	3	1
	Col Max	2	3	2 is solution

(Colin; Min of max min)

In this game, Rose plays R1 and Colin plays C1. Both movement arrows and saddle point method confirm that 2 at R1C1 is the pure strategy solution.

Example 5.5: Two-Person with Four Strategies for each Player

		C1	C2	C3	C4	
	R1	4	5	5	8	4
Rose	R2	6	7	6	9	6 Max
	R3	5	7	5	4	4
	R4	6	6	5	5	5
		6 min	7	6 min	9	Tie solution {6}

(Colin)

Does row min (max) == Column max (min)? Yes, we have multiple pure strategy solutions. Rose plays either R2 or R4, Colin plays either C1 or C3, and the solution to the game for Rose is 6.

5.3 Dominance and Dominated Strategies

Dominance and dominated strategies are helpful to attempt to reduce the size of a matrix game and even solve a game.

We start by defining the concept of dominance.

Row R_i dominates if all values in row R_i are *greater* than or equal to the corresponding payoff in another row.

Column C_i dominates if all values in column C_i are *smaller* than or equal to all values in the corresponding column.

The purpose of domination is to see if we can obtain a smaller payoff matrix.

Example 5.6: Dominance Application

	C1	C2	C3
R1	0	−1	1
R2	0	0	2
R3	−1	−2	3

Column C2 dominates both Columns C1 and C3. We are left with only Column C2

	C2
R1	−1
R2	0
R3	−2

Row R2 dominates rows R1 and R3.
Solution is 0.
In this example, dominance provided a solution to the game.

Example 5.7: Predator-Prey

In the following example, there is neither row nor column dominance,

Predator	Ambush	Pursue
Prey: Hide	0.2	0.4
Run	0.8	0.6

Example 5.8: Sports College Football

The row player is the offense and the column player is the defense. The values represent the yards gained (+) or lost (−) for plays R1–R5 versus defenses C1–C3.

Team A/Team B	C1	C2	C3
R1	0	−1	5
R2	7	5	10
R3	15	−4	−5
R4	5	0	10
R5	−5	−10	10

Dominance: C2 dominates C1.
Now, we reduce the game to the following payoff matrix

	C2	C3
R1	−1	5
R2	5	10
R3	−4	−5
R4	0	10
R5	−10	10

Row R2 dominates all other rows and we are left with

5	10

Column 2 dominates Column 3.
Five is the solution to the game.

Example 5.9: Consider the Following Game

		Colin			
		C1	C2	C3	Row Min
Rose	R1	1	1	10	1
	R2	2	3	−4	−4
Col Max		2	3	10	No saddle point solution

What now? Let's look for dominance,
Column C1 dominates Column C2
We are left with

		Colin	
		C1	C3
Rose	R1	1	10
	R2	2	−4

Now what? The game is reduced and still no pure strategy equilibrium.
We will eventually use a mixed strategy to solve this reduced-size game.

Chapter 5 Exercises: Section 5.1 Pure Strategy Games

Using movement diagram and/or dominance determine the outcomes of each game.

5.1

Attack and Defend Tableau		**Colonel Blotto**	
		Defend City I	**Defend City II**
Colonel Sotto	Attack City I	10	20
	Attack City II	0	0

What should Colonel Blotto and Colonel Sotto do?

5.2

Payoff Matrix		**Colin**	
		C1	C2
Rose	R1	a	b
	R2	c	d

What assumptions about the values {a b, c, d} have to be true for a at R1C1 to be the pure strategy solution?

5.3

Payoff Matrix		**Colin**	
		C1	C2
Rose	R1	a	b
	R2	c	d

What assumptions about the values {a, b, c, d} have to be true for b at R1C2 to be the pure strategy solution?

5.4

Payoff Matrix		**Colin**	
		C1	C2
Rose	R1	a	b
	R2	c	d

What assumptions about the values {a, b, c, d} have to be true for *c* at R2C1 to be the pure strategy solution?

5.5

Payoff Matrix		Colin	
		C1	C2
Rose	R1	a	b
	R2	c	d

What assumptions about the values {a, b, c, d} have to be true for *d* at R2C2 to be the pure strategy solution?

5.6

		Colin	
		C1	C2
Rose	R1	0.5	0.5
	R2	2	0

What is the pure strategy solution for this game?

5.7 Consider the following batter-pitcher duel. All the entries in the payoff matrix reflect the percent of hits off the pitcher, the batting average. What strategies are optimal for each player?

Payoff Tableau		Pitcher	
		Fastball C1	Knuckleball C2
Batter	Guess Fastball R1	.425	.125
	Guess Knuckleball R2	.325	.275

5.8 Consider the following game, find the solution.

Rose/Colin	C1	C2	C3
R1	80	40	75
R2	70	35	30

5.9 Consider the following game, find the solution.

Rose/Colin	C1	C2	C3	C4
R1	40	80	35	60
R2	65	90	55	70
R3	55	40	45	75
R4	45	25	50	50

5.10 The following represents a game between a professional athlete (Rose) and management (Colin). The values are in thousands. What decision should each make?

Rose/Colin	C1	C2	C3
R1	490	220	195
R2	425	350	150

5.11 Solve the following game.

| | | Colin | |
		C1	C2
Rose	R1	(2,−2)	(1,−1)
	R2	(3,−3)	(4,−4)

5.12 Solve the following game.

| | | Colin | |
		C1	C2
Rose	R1	2	1
	R2	3	4

5.13 Given the following game:

| | | Colin | | |
		C1	C2	C3
Rose	R1	4	4	10
	R2	2	3	1
	R3	6	5	7

5.14 Solve the following network TV dilemma:

| | | Network 2 | | |
		Crime Drama	Reality	Comedy
	Crime Drama	35	15	60
Network I	Reality	45	55	50
	Comedy	38	14	70

5.15 Solve the following game:

| | | Colin | | |
		C1	C2	C3
Rose	R1	0	−1	1
	R2	0	0	2
	R3	−1	−2	3

5.16 The predator has two strategies for catching the prey (ambush or pursuit). The prey has two strategies for escaping (hide or run). The game matrix

| Payoff Tableau | | Predator | |
| | | *Ambush* | *Pursue* |
		C1	*C2*
Prey	Hide R1	.20	.40
	Run R2	.80	.60

5.17 A professional football team has collected data for certain plays against certain defenses. In the payoff matrix, the values are the yards gained or lost for a particular play against a particular defense.

Team A/Team B	C1	C2	C3
R1	0	−1	5
R2	7	5	10
R3	15	−4	−5
R4	5	0	10
R5	−5	−10	10

5.4 Mixed Strategy in Two-Player Two-Strategy Games

A **mixed strategy** is an assignment of a probability to each pure strategy. It defines a probability over the strategies, and reflects that, rather than choosing a particular pure strategy, the player will randomly select a pure strategy based on the distribution given by their mixed strategy. Of course, every pure strategy is a mixed strategy which selects that particular pure strategy with probability 1 and every other strategy with probability 0.

Example 5.10: Pitcher-Batter Game with Payoff as Batting Averages

Batter Guesses	Pitcher Pitches	
	Fastball (FB)	Curveball (CB)
Fastball (FB)	.300	.200
Curveball(CB)	.100	.500

There is no saddle point nor any dominant solution.

We must employ the use of mixed strategies to find a solution. Let FB be fastball and CB be curveball.

Method 1: Equating Expected Value and Using Algebra

For the batter, let p = probability that the batter guesses FB and $q = (1 - p)$ = probability that the batter guesses curve. We set the expected values equal and solve.

$$.3p + .1(1-p) = .2p + .5(1-p)$$
$$.2p + .1 = -.3p + .5$$
$$.5p = .4$$
$$p = 0.8$$
$$q = (1-p) = 0.2$$

The expected payoff is .260.

Now for the pitcher, let x = probability that they throw a FB and $1 - x = y =$ probability that they throw a CB.

Now,

$$.3x + .2(1 - x) = .1x + .5(1 - x)$$

$$.1x + .2 = -.4x + .5$$

$$.5x = .3$$

$$x = 0.6 \text{ and } (1 - x) = y = 0.4$$

The expected payoff is 0.260.

The batter can get his best results by guessing FB 80% of the time and CB 20%, while the pitcher gets his best results by throwing the FB 60% of the time and CB 40% of the time. The batter can do no better than 0.260.

Method of Oddments (Known also as William's Oddments)

Oddments is a shortcut for 2 × 2 games to equate Expected Value.

In oddments, for each row or column calculated, the subtraction is done so that the result is always non-negative. Note in the example below, in the first row, we perform an operation 0.3 − 0.2 to get 0.1, and in the second row, we take 0.5 − 0.1 to get 0.4.

Batter Guesses	Pitcher Pitches			
	Fastball (FB)	Curveball (CB)	Oddments	Swap and Divide
Fastball (FB)	.300	.200	.3 − .2 = .1	0.4/0.5 = .8
Curveball (CB)	.100	.500	.5 − .1 = .4	.1/.5 = .2
Oddments	.3 − .1 = .2	.5 − .2 = .3	Sums are 0.5	
Swap and Divide	.3/.5 = .6	.2/.5 = .4		

Note we first subtract (keeping the results positive) then switch positions, and divide by the total to get the probabilities we desire. Thus, the batter guesses FB 80% of the time and CB 20% of the time, while the pitcher actually pitches an FB 60% of the time and CB 40% of the time. The batter will hit 0.260 as an average using this method. The calculations for the expected average are:

Expected value is found by multiplying payoffs by respective probabilities. They should all be equal.

$$.8 * .3 + .2 * .1 = .260$$

$$.8 * .2 + .2 * .5 = .260$$

$$.3 * .6 + .2 * .4 = .260$$

$$.1 * .6 + .5 * .4 = .260$$

Note: For 2 × 2 games with mixed strategy solutions, we find this method the easiest.

In the next few examples, we extend William's method for 2 × n or m × 2 games.

William's Method: For 2 × n or n × 2 games.

Example 5.11: Williams Method for the Game Below

		Colin				
		C1	C2	C3	C4	C5
Rose	R1	−2	5	1	0	−4
	R2	3	−3	−1	3	8

There is no saddle point solution and no observed dominance so there is a mixed strategy solution. Here is what we do.

1. Choose the player that has only two strategies and we plot the points from R1 to R2 for each of C1–C5. Since Rose uses the *Maximin criterion*, we use the lower portion of the graph (min) and choose the highest point on the lower portion.
2. We then use the points that form that intersection and use them as reduced payoff matrix.
3. Then we will use oddments to get a solution.
4. Apply the solution to the original problem.

The intersection of C1 and C3 from the intersection that represents the Max of the Min portion of the graph is shown in Figure 5.1. Thus, we reduce the payoff matrix to:

		Colin	
		C1	C3
Rose	R1	−2	1
	R2	3	−1

By oddments we find

		Colin		Oddments	
		C1	C3		
Rose	R1	−2	1	3	4/7
	R2	3	−1	4	3/7
Oddments		5	2		
		2/7	5/7		V = 1/7

Overall, Colin plays C1 (2/7), C2 (0), C3 (5/7), C4 (0), and C5 (0), while Rose plays R1 (4/7) and R2 (3/7) to obtain their best strategies and solution of 1/7.

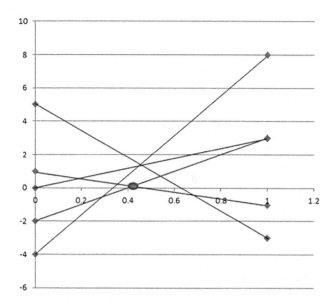

FIGURE 5.1
William's method applied to Example 5.11.

Example 5.12 A 3 × 2 Game with William's Method

		Colin	
		C1	C2
Rose	R1	2	−3
	R2	0	2
	R3	−5	10

Again there is no saddle solution or dominance. We must use mixed strategies.

Since this is a 3 × 2 game. We can plot Rose's three versus Colin's two. Colin wants to minimize his maximum result. In this case, we take the upper portion of the graph's intersections and find its lowest point.

We find that R1 intersecting with R2 determines the Min of the Max portion of the graph. We eliminate R3 (stating we will never play it). We are left with the following payoff matrix (Figure 5.2):

		Colin		Oddments	
		C1	C2		
Rose	R1	2	−3	5	2/7
	R2	0	2	2	5/7
Oddments		2	5		
		5/7	2/7		V = 4/7

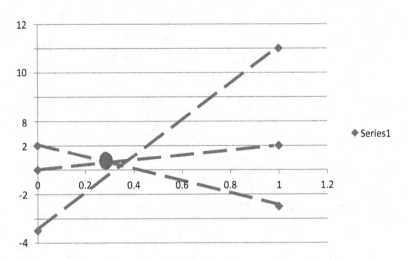

FIGURE 5.2
William graph for Example 5.12.

Chapter 5 Exercises: Section 5.2 Mixed Strategy Games

5.18

Payoff Tableau		Colin	
		C1	C2
Rose	R1	a	b
	R2	c	d

What assumptions have to be true for there not to be a saddle point solution? Show that the two largest entries must be diagonally opposite each other.

5.19 Given

Payoff Tableau		Colin	
		C1	C2
Rose	R1	a	b
	R2	c	d

where a > d > b > c. Show that Colin play C1 and C2 with probabilities x and $(1 - x)$ that

$$x = \frac{d-b}{(a-c)+(d-b)}.$$

5.20 In the game, in Example 5.19, show the value of the game is

$$v = \frac{ad - bc}{(a-c)+(d-b)}.$$

5.21 Solve the game.

Payoff Tableau		Colin	
		C1	C2
Rose	R1	6	4
	R2	4	2

5.22 Solve the game.

Payoff Tableau		Colin	
		C1	C2
Rose	R1	-2	3
	R2	2	-2

5.23 Solve the hitter-pitcher duel for the following players:

Payoff Tableau		RR	
		Fastball C1	Split finger C2
Derek Jeter	Guess Fastball R1	.330	.250
	Guess split finger R2	.180	.410

5.24 Solve the game.

Payoff Tableau		Colin		
		C1	C2	C3
Rose	R1	3	7	2
	R2	8	5	1
	R3	6	9	4

5.25 Solve the game.

Payoff Tableau		Colin		
		C1	C2	C3
Rose	R1	0.5	0.9	0.9
	R2	.1	0	.1
	R3	.9	.9	.5

5.26 Solve the game.

Payoff Tableau		Colin	
		C1	C2
Rose	R1	6	5
	R2	1	4
	R3	8	5

5.27 Solve the game.

Payoff Matrix		Colin	
		C1	C2
Rose	R1	2	−1
	R2	1	4
	R3	6	2

5.28 Solve the game.

		Colin			
		C1	C2	C3	C4
Rose	R1	1	−1	2	3
	R2	2	4	0	5

5.29 Solve the following pitcher-hitter game.

		Pitcher	
		Fastball	Curveball
Batter	Fastball	.300	.200
	Curveball	.100	.500

5.30 Solve the game.

		Colin		
		A	B	Row Min
Rose	A	2	−3	−3
	B	0	2	0
	C	−5	10	−5
Column Max		−2	10	

5.31 Solve the game.

	#1	#2	#3
#1	1	1	10
#2	2	3	-4

5.32 Solve the game.

		Colin		
		A	B	C
Rose	A	1	2	2
	B	2	1	2
	C	2	2	0

5.5 Linear Programming and Total Conflict Games

Linear programming can ALWAYS be used to find the value of the game and the optimal strategies (for Rose and Colin) for any two-person zero-sum or constant-sum game. Our objective here is to present a "how to" approach to convert a zero-sum game to a linear program and provide several examples. Technology is essential to solve linear program games. EXCEL, LINDO, LINGO, Gambit, and Python may be used. We will illustrate several of these LP solvers. The key is the formulation of the linear program.

5.5.1 Formulations of Game Theory Payoff Matric into Linear Programming

Given a payoff matrix with just Rose's values such as below

		Colin			
		C_1	C_2	...	C_m
	R_1	$a_{1,1}$	$a_{1,2}$	...	$a_{1,m}$
Rose	R_2	$a_{2,1}$	$a_{2,2}$	...	$a_{2,m}$
	...	...	...	...	...
	R_n	$a_{n,1}$	$a_{n,2}$	...	$a_{n,m}$

Case 1. All a_{ij} are ≥ 0.

For Rose, the decision variables are the probabilities associated with R1, R2, R3, ... Rn, call them x1, x2, x3, ... xn. The objective function is to maximize VR, the value of the game to Rose.

The constraints are all functions of the decision variables, the coefficients, a_{ij} and the value of the game, VR.

The LP formulation is

Maximize V_R

Subject to:
$$a_{1,1}x_1 + a_{2,1}x_2 + \cdots + a_{n,1}x_n - V_R \geq 0$$
$$a_{1,2}x_1 + a_{2,2}x_2 + \cdots + a_{2,n}x_n - V_R \geq 0$$
$$\cdots$$
$$a_{1,m}x_1 + a_{2,m}x_2 + \cdots + a_{m,n}x_n - V_R \geq 0$$
$$x_1 + x_2 + \cdots x_n = 1$$
$$x_i \geq 0, i = 1, 2, \ldots n$$

We solve the LP using a LP solver (LINDO or EXCEL) to solve and examine the LP's shadow prices (negative dual solution) (directly or through sensitivity analysis) to get the probabilities associated with Colin's decisions, y1, y2, y3, ..., ym.

Case 2. All a_{ij} are <u>not greater than or equal to</u> 0

For Rose, the decision variables are the probabilities associated with R1, R2, R3, ..., Rn, call them x1, x2, x3, ..., xn. The objective function is to maximize VR, the value of the game to Rose.

The constraints are all functions of the decision variables, the coefficients, a_{ij} and the value of the game, VR.

Since some $a_{ij} < 0$ then the value of the game might be negative. We replace V_R with the difference of two decision variables that are both ≥ 0. Let $V_R = V_{R1} - V_{R2}$

The LP formulation would be:

Maximize $V_{R1} - V_{R2}$

Subject to:
$$a_{1,1}x_1 + a_{2,1}x_2 + \cdots + a_{n,1}x_n - V_{R1} + V_{R2} \geq 0$$
$$a_{1,2}x_1 + a_{2,2}x_2 + \cdots + a_{2,n}x_n - V_{R1} + V_{R2} \geq 0$$
$$\cdots$$
$$a_{1,m}x_1 + a_{2,m}x_2 + \cdots + a_{m,n}x_n - V_{R1} + V_{R2} \geq 0$$
$$x_1 + x_2 + \cdots + x_n = 1$$
$$x_i \geq 0, i = 1,2,\ldots n V_{R1}, V_{R2} \geq 0$$
$$V_{Ri} \geq 0 i = 1,2$$

Let's illustrate the linear programming methods with a few examples.

Example 5.13: Pure Strategy Solution

		Colin	
		C1	C2
Rose	R1	5	1
	R2	3	0

The formulation is:

Maximize V

Subject to:
$$5x_1 + 3x_2 - V \geq 0$$
$$x_1 - V \geq 0$$
$$x_1 + x_2 = 1$$
$$x_1, x_2, V \geq 0$$

Using the Solver in Excel as discussed earlier in Chapter 3, we obtain that the objective function value is 1 when $x_1 = 1$ and $x_2 = 0$. From the shadow prices, which represent Player 2's solution, we find $y_1 = 0$ and $y_2 = 1$.

Next, we solve using *LINDO. The formulation to be put in LINDO is*

max V
$5 \times 1 + 3 \times 2 - V > 0$
$X1 - V > 0$
$x1 + x2 + 1$
end

LP OPTIMUM FOUND AT STEP   2
   OBJECTIVE FUNCTION VALUE
   1)   1.000000

VARIABLE	VALUE	REDUCED COST
V	1.000000	0.000000
X1	1.000000	0.000000
X2	0.000000	1.000000

ROW	SLACK OR SURPLUS	DUAL PRICES
2)	4.000000	0.000000
3)	0.000000	−1.000000
4)	0.000000	1.000000

NO. ITERATIONS=   2
RANGES IN WHICH THE BASIS IS UNCHANGED:

OBJ COEFFICIENT RANGES

VARIABLE	CURRENT COEF	ALLOWABLE INCREASE	ALLOWABLE DECREASE
V	1.000000	INFINITY	1.000000
X1	0.000000	INFINITY	1.000000
X2	0.000000	1.000000	INFINITY

RIGHTHAND SIDE RANGES

ROW	CURRENT RHS	ALLOWABLE INCREASE	ALLOWABLE DECREASE
2	0.000000	4.000000	INFINITY
3	0.000000	1.000000	4.000000
4	1.000000	INFINITY	1.000000

The solution is that the value of the game is (1, −1) when Rose always plays R1 ($x1 = 1$) and Colin always plays C2 ($y2 = 1$).

Example 5.14: A Mixed Strategy Solution with Linear Programming

		Colin	
		C1	C2
Rose	R1	1	3
	R2	5	2

Maximize V_R

Subject to:
$$x_1 + 5x_2 - V_R \geq 0$$
$$3x_1 + 2x_2 - V_R \geq 0$$
$$x_1 + x_2 = 1$$
$$x_1, x_2, V_R \geq 0$$

Using technology, we obtain that the objective function value is 2.6 when $x_1 = 0.6$ and $x_2 = 0.4$. From the shadow prices, which represent the *negative of the dual solution*, we find that $y_1 = 0.2$ and $y_2 = 0.8$.

Using LINDO,

LP OPTIMUM FOUND AT STEP   2
    OBJECTIVE FUNCTION VALUE
      1)   2.600000

VARIABLE	VALUE	REDUCED COST
VR	2.600000	0.000000
X1	0.600000	0.000000
X2	0.400000	0.000000

ROW	SLACK OR SURPLUS	DUAL PRICES
2)	0.000000	−0.200000
3)	0.000000	−0.800000
4)	0.000000	2.600000

NO. ITERATIONS=   2
RANGES IN WHICH THE BASIS IS UNCHANGED:

OBJ COEFFICIENT RANGES

VARIABLE	CURRENT COEF	ALLOWABLE INCREASE	ALLOWABLE DECREASE
VR	1.000000	INFINITY	1.000000
X1	0.000000	4.000000	1.000000
X2	0.000000	1.000000	4.000000

RIGHTHAND SIDE RANGES

ROW	CURRENT RHS	ALLOWABLE INCREASE	ALLOWABLE DECREASE
2	0.000000	3.000000	2.000000
3	0.000000	2.000000	3.000000
4	1.000000	INFINITY	1.000000

Example 5.15: 2 × 3 Game with Negative Entries in the Payoff Matrix

		Colin	
Rose	C1	C2	C3
R1	1	−1	3
R2	−2	1	−1

Note, some entries are negative so we replace VR with $VR1 - VR2$ in the objective function. Our formulation is from the perspective of Rose as:

Maximize $Z = VR1 - VR2$

Subject to:
$x1 - 2 \times 2 - VR1 + VR2 \geq 0$
$-x1 + x2 - VR1 + VR2 \geq 0$
$3x1 - x2 - VR1 + VR2 \geq 0$
$x1 + x2 = 1$
all variables ≥ 0

The solution for Rose, $Vr = -0.2$ when Rose plays 0.6 $R1$ and 0.4 $R2$. What about Colin?

The shadow prices are the values to Colin: value is 0.2 when $c_1 = 0.4$, $c_2 = 0.6$, and $c_3 = 0$.

Colin never plays c_3.

In LINDO, notice how we had to rewrite constraint 2,

Maximize VR1 − VR2

Subject to:
$x1 - 2x2 - VR1 + VR2 > 0$
$x2 - x1 - VR1 + VR2 > 0$
$3x1 - x2 - VR1 + VR2 > 0$
$x1 + x2 = 1$
end

LP OPTIMUM FOUND AT STEP   2
LP OPTIMUM FOUND AT STEP   1
    OBJECTIVE FUNCTION VALUE
    1)   −0.2000000

VARIABLE	VALUE	REDUCED COST
VR1	0.000000	0.000000
VR2	0.200000	0.000000
X1	0.600000	0.000000
X2	0.400000	0.000000

ROW	SLACK OR SURPLUS	DUAL PRICES
2)	0.000000	−0.400000
3)	0.000000	−0.600000
4)	1.600000	0.000000
5)	0.000000	−0.200000

NO. ITERATIONS=   1

Again, we look at the shadow prices to obtain that Colin plays 0.4 C1, 0.6 C2, and 0.0 C3. *The value of the game to Colin is Vc = 0.2.*

Example 5.16: A 3 × 3 Example where Equalizing Strategies Algebra Methods do not Work

		Colin			
		C1	C2	C3	
	R1	9	2	7	x
Rose	R2	3	6	4	y
	R3	5	3	1	$1 - x - y$
		w	z	$1 - w - z$	

Go ahead and try equalizing strategies to find the mix of strategies to play to obtain the value of the game. You will find that the method does not work. Since equalizing strategies did not work, the problem is a 3 × 3; one method is to solve all the subgames (2 × 3, 2 × 2, etc.). We suggested solving every 2 × 3 subgame using the graphical method. There are three subgames from Rose and three subgames for Colin. This involves solving six games and then trying to figure out the solution from the results. It turns out in this example that the games that match strategies in the mix with the highest expected value are the solution.

However, linear programming works. Linear programming techniques work whether or not the solution has saddle point or mixed strategies. In other words, it always works.

For Rose,

Decision Variables:
    v = max expected value
    x = probability for playing strategy A
    y = probability for playing strategy B
    z = probability for playing strategy C

Objective Function;

Maximize v

Subject to:

$9x + 3y + 5z - v \geq 0$

$2x + 6y + 3z - v \geq 0$

$7x + 4y + 1z - v \geq 0$

$x + y + z = 1$

nonnegativity x, y, z, v $\geq$ 0

Again using technology, we obtain the solution as $v = 4.8$ when $x = 0.3$, $y = 0.7$, and $1 - x - y = 0$. From the shadow prices, we obtain the solution for Colin as $w = 0.6$, $z = 0.6$, and $1 - w - z = 0.0$, and the value for Colin from the two-sum game is −4.8.

In LINDO, LP OPTIMUM FOUND AT STEP   3

   OBJECTIVE FUNCTION VALUE

   1)   4.800000

VARIABLE	VALUE	REDUCED COST
V	4.800000	0.000000
X	0.300000	0.000000
Y	0.700000	0.000000
Z	0.000000	1.000000

ROW	SLACK OR SURPLUS	DUAL PRICES
2)	0.000000	−0.400000
3)	0.000000	−0.600000
4)	0.100000	0.000000
5)	0.000000	4.800000

NO. ITERATIONS=   3

RANGES IN WHICH THE BASIS IS UNCHANGED:

OBJ COEFFICIENT RANGES

VARIABLE	CURRENT COEF	ALLOWABLE INCREASE	ALLOWABLE DECREASE
V	1.000000	INFINITY	1.000000
X	0.000000	4.000000	2.000000
Y	0.000000	6.000000	2.000000
Z	0.000000	1.000000	INFINITY

RIGHTHAND SIDE RANGES

ROW	CURRENT	ALLOWABLE	ALLOWABLE
	RHS	INCREASE	DECREASE
2	0.000000	7.000000	0.142857
3	0.000000	3.000000	0.333333
4	0.000000	0.100000	INFINITY
5	1.000000	INFINITY	1.000000

Example 5.17: Rock, Paper, Scissor Game

The payoff matrix for winning worth 1 and losing worth −1 would be:

Payoff Matrix		Colin		
		Rock	Paper	Scissor
Rose	Rock	0	−1	1
	Paper	1	0	−1
	Scissor	−1	1	0

What should each player do to optimize their ability to win the game? Appling linear programming methods, our formulation with Rose's options {x1, x2, x3} and Colin options {y1, y2, y3} and noticing that there are both positive and negative outcomes (we use VR = VR1 − VR2, where both VR1 and VR2 ≥ 0)

Max VR1 − VR2

Subject to:
$$x2 - x3 - (VR1 - VR2) \geq 0$$
$$-x1 + x3 - (VR1 - VR2) \geq 0$$
$$x1 + x2 - (VR1 - VR2) \geq 0$$
$$x1 + x2 + x3 = 2$$
$$x1, x2, x3, VR1, VR2 \geq 0$$

The solution is for Rose to play x1 = x2 = x3 = 1/3 with the value to Rose equaling 0 and Colin plays y1 = y2 = y3 = 1/3 and the value to Colin is 0.

So, we know how to play the game, but how do we implement it? First, we must decide to play the game many times. Second, we must develop a random scheme to play the Rock, Paper, and Scissors each 1/3 of the time in a random fashion.

We illustrate using Gambit. Gambit is a software package for solving game theory problems and is a free download.

Let's return to the rock-paper-scissor game with Rock beats Scissors, Scissors beat Papers, and Paper beats Rock.

Rock-Paper-Scissors

We enter a 1 if one choice beats another, a −1 if one choice loses to another, and a 0 if they tie.

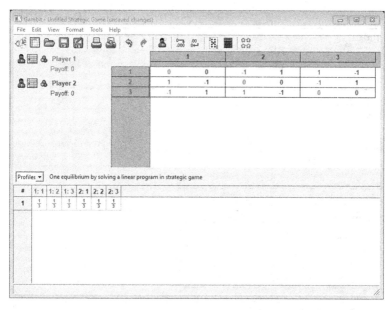

We solve using Gambit and found the solution to be (0, 0) when each player plays Rock, Paper, and Scissors 1/3 of the time. Obviously, we have to play this game many times to break even.

Chapter 5 Exercises: Section 5.3 LP Games

Solve using Linear Programming (LP). Consider the following zero-sum games, where the payoffs represent gains to row player (1) and losses to the column player (2). Solve each game.

5.32

Payoff Tableau		Colin	
		C1	C2
Rose	R1	6	4
	R2	4	2

5.33

Payoff Tableau		Colin	
		C1	C2
Rose	R1	−2	3
	R2	2	−2

5.34 Solve the hitter-pitcher duel for the following players:

Payoff Tableau		Verlander	
		Fastball C1	Split finger C2
Judge	Guess Fastball R1	.330	.250
	Guess split finger R2	.180	.410

5.35

Payoff Tableau		Colin		
		C1	C2	C3
Rose	R1	3	7	2
	R2	8	5	1
	R3	6	9	4

5.36

Payoff Tableau		Colin		
		C1	C2	C3
Rose	R1	0.5	0.9	0.9
	R2	.1	0	.1
	R3	.9	.9	.5

5.37

Payoff Tableau		Colin	
		C1	C2
Rose	R1	6	5
	R2	1	4
	R3	8	5

5.38

Payoff Tableau		Colin	
		C1	C2
Rose	R1	2	−1
	R2	1	4
	R3	6	2

5.39

		C1	C2	C3	C4
			Colin		
Rose	R1	1	−1	2	3
	R2	2	4	0	5

5.40 Solve the pitcher-hitter game.

		Pitcher	
		Fastball	Curveball
Batter	Fastball	.300	.200
	Curveball	.100	.500

5.41 Solve the game.

		C1	C2	C3	C4	C5
				Colin		
Rose	R1	.5	.5	.5	1	1
	R2	1	.5	.5	.5	1
	R3	1	1	.5	.5	.5
	R4	1	1	1	.5	0

5.42

		C1	C2	C3
			Colin	
Rose	R1	1	1	10
	R2	2	3	−4

5.43

		C1	C2	C3
			Colin	
Rose	R1	1	2	2
	R2	2	1	2
	R3	2	2	0

Summary

In this chapter, we discuss total conflict simultaneous games, also known as zero-sum and constant-sum games between two players. We illustrated the Nash equilibrium for two types of solutions: pure strategy and mixed strategy. As a Nash equilibrium for total conflict games, they represent the solution to the game.

In a plot of the entries in a total conflict game, Figure 5.3, we see for this example the coordinates lie in a straight line. This will be true for all total conflict games.

		Colin	
		C1	**C2**
Rose	R1	(2, −2)	(1,−1)
	R2	(−4, 4)	(0, 0)

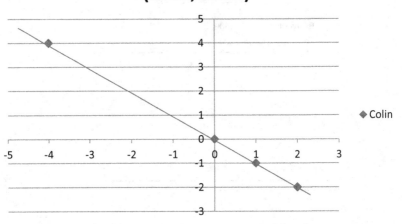

(Rose, Colin)

FIGURE 5.3
Plot of entries for (Rose, Colin) in a total conflict game showing the linear relationship.

Reference

Nash, J. (1950). Equilibrium points in n-person games. *Proceeding of the National Academy of Sciences*, (36), 48–49.

6

Partial Conflict Games: The Classical Two-Player Games

6.1 Partial Conflict Simultaneous Games Introduction

Given a game such as:

		Colin			
		C1	C2	...	Cn
Rose	R1	(m_1, n_1)	(m_1, n_2)		(m_1, n_n)
	R2	(m_2, n_1)	(m_2, n_2)		(m_2, n_n)
	...				
	Rm	(m_m, n_1)	(m_m, n_2)	...	(m_m, n_n)

In this payoff matrix, the sum of the entries (m, n_j) for $i = 1, 2, \ldots m$ and $j = 1, 2, \ldots n$ does not all sum to 0 nor all sum to the same constant. Here the payoff to one player is not exactly the opposite of the other players. Both can win or lose something, or tie.

As in zero-sum games, all these partial conflict games have solutions. The solution can be either pure strategy or mixed (equalizing) strategies. We discuss pure strategies in this chapter and equalizing strategies in Chapter 8.

As in total conflict games, we can use movement diagrams to look for pure strategy Nash equilibriums. First, we define a Nash equilibrium again.

DEFINITION OF NASH EQUILIBRIUM

When no player can benefit by departing unilaterally (by itself) from its strategy associated with an outcome, the strategies constitute a **Nash equilibrium**.

DOI: 10.1201/9781032726885-6

MOVEMENT DIAGRAMS

MOVEMENT DIAGRAMS in non-zero-sum games. For <u>Rose</u>, she would maximize payoffs, so she would prefer the highest payoff at each <u>column</u>. Arrows in columns but values are Rose's. Similarly, for Colin, he wants to maximize his payoffs, so he would prefer the high payoff at each row. We draw an arrow to the highest payoff in that row. Arrows are in rows but values are Colin's. If all arrows point in to an entry from every direction, then that or those points will be pure Nash equilibrium.

The Nash equilibrium can be either in pure or mixed strategies. Every 2×2 non-zero-sum game has at least one Nash equilibrium. However, in contrast to the total conflict games discussed in Chapter 5, the Nash equilibrium may not be the solution to the game.

We will begin our discussion with simultaneous games.

Possible methods to play the game:

1. **Maximize his payoffs.** Each player chooses a strategy in an attempt to maximize his payoff. While he reasons what the other player's response will be, he does not have the objective of ensuring the other player gets a "fair" outcome. Instead, he "selfishly" maximizes his payoff.

2. **Find a stable outcome.** Quite often players have an interest in finding a stable outcome. *A Nash equilibrium outcome is an outcome from which neither player can unilaterally improve* and therefore represents a stable situation. For example, we may be interested in determining whether two species in a habitat will find equilibrium and coexist, or will one species dominate and drive the other to extinction? The Nash equilibrium is named in honor of John Nash who proved (Nash, 1950) that every two-person game has at least one equilibrium in either pure or mixed strategies.

3. **Minimize the opposing player.** Suppose we have two corporations whose marketing of products interact with each other, but not in total conflict. Each may begin with the objective of maximizing his payoffs. But, if dissatisfied with the outcome, one, or both corporations, may turn hostile and choose the objective of minimizing the other player. That is, a player may forego their long-term goal of maximizing their own profits and choose the short-term goal of minimizing the opposing player's profits. For example, consider a large, successful corporation attempting to bankrupt a "start-up venture" in order to drive him out of business, or perhaps motivate him to agree to an arbitrated "fair" solution.

4. **Find a "mutually fair" outcome, perhaps with the aid of an arbiter.**
Both players may be dissatisfied with the current situation. Perhaps, both have a poor outcome as a result of minimizing each other. Or perhaps one has executed a "threat' as we study below, causing both players to suffer. In such cases, the players may agree to abide by the decision of an arbiter who must then determine a "fair" solution (Nash, 1950).

We only present the stable outcome in this chapter.
We will begin our study of partial conflict games with a few classical games.

6.2 The Prisoner's Dilemma

We start by formulating the game known as the Prisoner's Dilemma, modeling an example where Country A and Country B could either Arm or Disarm:

	Country B	
	Disarm	**Arm**
Disarm	(3, 3)	(1, 4)
Country A		
Arm	(4, 1)	(2, 2)

In Figure 6.1, we note that a plot of the payoffs to each player does not lie on a line, indicating that the game is not in total conflict. We can also note

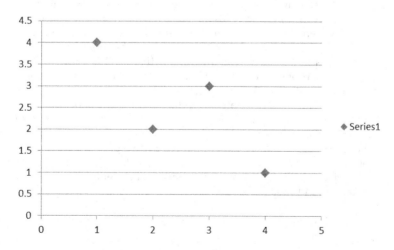

FIGURE 6.1
Payoffs in a partial conflict game do not lie on a line.

the sum of the payoffs if both players' Arm is 2 + 2 = 4, while the sum of the payoffs if both Disarm is 3 + 3 = 6, not a constant sum. Note that if the players can move from (2, 2) to (3, 3), both players improve, which is not possible in total conflict.

What are the objectives of the players in a partial conflict game? In total conflict, each player attempts to maximize his payoffs while minimizing the other player in the process. But in a partial conflict game, a player may have any of the following objectives that we have mentioned but we only discuss the stable outcomes.

Example 6.1: Prisoner's Dilemma Without Communication

We will assume that without communication each player will play "conservatively". That is, each player will play his maximin strategy. Country A's payoffs are listed first, with the ranking of 4 being best.

	Country B		Player A: Row Minimum
	Disarm	Arm	
Disarm	(3, 3) ⟹	(1, 4)	1
Country A	↓	↓	
Arm	(4, 1) ⟹	(2, 2)	2
Player B: Column Minimum	1	2	

We see that Country A's maximin strategy is Arm since he is guaranteed at least a 2 if he plays Arm. Similarly, Country B's maximin strategy is Arm. If both players play conservatively without communication, we would expect the result to be (Arm, Arm) with payoff (2, 2). Also note from the movement diagram that (2, 2) is the Nash equilibrium outcome, so neither player unilaterally can improve and that both players have a dominant strategy, Arm. The dilemma is that (3, 3) is better for both players. In the problem set, you are asked to show that only a promise by either player will yield (3, 3). We can generalize the Prisoner's Dilemma game by considering the strategies to be Defect and Cooperate. In the above application, Arm is Defect and Disarm is Cooperate. The term Prisoner's Dilemma was coined by Princeton mathematician Albert Tucker in 1950 to describe a situation faced by two prisoners held in isolation and both can Squeal (Defect) or Not Squeal (Cooperate with their fellow prisoner). The situation is described in Problems 10.6 after communications and cooperation are allowed.

Summarizing the game of Prisoner's Dilemma

Prisoner's Dilemma is a two-person partial conflict game in which each player has two strategies, Defect or Cooperate, Defect dominates Cooperate for both players even though the mutual defection outcome, which is the unique Nash equilibrium in the game, is worse for both players than the mutual cooperation outcome.

The Prisoner's Dilemma is widely used to model applications such as global warming, cigarette advertising, drug addiction, evolutionary biology, and oil companies' pricing strategies.

6.3 The Game of Chicken

Two cars drive straight at each other. The first driver to swerve loses the duel. Each player has two options, Swerve and Not Swerve. The worst outcome is if both players choose Not Swerve causing a crash. The best outcome for each player is to win the game; that is, they execute Not Swerve, and the opposing player Swerve. The next best outcome for each player is if both players Swerve, resulting in a tie. With the ranking of 4 being best and 1 worst, we could have the payoff matrix:

	Colin	
	Swerve	Not Swerve
Swerve	(3, 3)	(2, 4)
Rose		
Not Swerve	(4, 2)	(1, 1)

Example 6.2: The Game of Chicken Without Communication

We again assume that without communication each player will play his maximin strategy.

	Colin		Rose: Row Minimum
	Swerve	Not Swerve	
Swerve	(3, 3) $\Longrightarrow$	(2, 4)	2
Rose	$\downarrow$	$\uparrow$	
Not Swerve	(4, 2) $\Longleftarrow$	(1, 1)	1
Colin: Column Minimum	2	1	

We see that Rose's maximin strategy is Swerve since she is guaranteed at least a 2 if she plays Swerve. Similarly, Colin's maximin strategy is Swerve.

If both players play conservatively without communication, we would expect the result to be (Swerve, Swerve) with payoff (3, 3). Note from the movement diagram that neither player has a dominant strategy and that (3, 3) is not a Nash equilibrium, and either player can improve unilaterally by switching to her or his Not Swerve strategy. However, if both players switch, then the result is a disastrous outcome (Not Swerve, Not Swerve) with payoffs (1, 1). There are two Nash equilibriums with payoffs (4, 2) and (2, 4). Now imagine a confrontation between two nations, such as the Cuba Missile Crisis discussed in Chapter 10. Both countries are sitting on Swerve, but have the motivation to switch to Not Swerve at the last minute. If both switch, a catastrophe occurs.

Summarizing the game of Chicken

Chicken is a two-person partial conflict game in which each player has two strategies: Swerve to avoid a crash or Not Swerve to attempt to win the game. Neither player has a dominant strategy. If both players choose Swerve, the outcome is not a Nash equilibrium and therefore unstable. There are two Nash equilibriums where one of the two players chooses Swerve and the second player chooses Not Swerve.

The game of chicken is used to model such topics as the confrontations between countries or species.

Chapter 6 Exercise

6.1 The following game is called the Battle of the Sexes. In this game, he wants to buy tickets for the baseball game, and she would like to buy tickets for the ballet.

		She buys a ticket for	
		Baseball game	Ballet
He buys a ticket for	Baseball game	(4, 3)	(2, 2)
	Ballet	(1, 1)	(3, 4)

Find the Nash equilibrium and discuss why it does not appear to be an optimal solution.

Reference and Further Readings

Nash, J. (1950). Equilibrium points in n-person games. *Proceeding of the National Academy of Sciences USA* (36), 48–49.
Straffin, P. (2004). *Game Theory and Strategy*. Washington, DC: MAA Press.

7

Utility Theory

7.1 Introduction

In most of our examples, we have assumed the values in the payoff matrix were given. We have not given much discussion, if any, to the values and their meaning. It is time to consider those numbers more thoroughly. We will discuss two types of numbers Ordinal and Cardinal. In order to do the kind of mathematics previously discussed we need cardinal numbers, where we can do mathematics. Ordinal Numbers review. Numbers are positions like first place, second place, etc. They could be numbers to record categorical information (red-1, blue-2, etc.).

Cardinal numbers ... why they are important in mixed strategies. Cardinal numbers are real numbers where the value has meaning: weight, height, age, income, number of casualties, etc.

Utilities are used to measure Preferences. Preferences are something like I like Diet Coke more than I like Diet Pepsi.

We will present some methods to obtain Utilities, Data, Analytical Hierarchy Process (AHP) or Technique of Order Preference by Similarity to Ideal Solution (TOPSIS), and Lotteries Methods by Morgenstern and Von Neumann.

We will find these methods useful to find expected utility to be used in place of the ordinal values used in Chapter 6.

7.2 Ordinal Numbers

Number are positions like first place, second place, etc. They could be numbers to record categorical information (red-1, blue-2, etc.). As categorical numbers, we cannot do mathematics. No addition, subtraction, multiplication, or division. It does not make sense to do mathematics.

Ordinal Numbers are those numbers that represent the precise position of an object in a particular space. If a number of things or people are given as a

DOI: 10.1201/9781032726885-7

list, ordinal numbers are used to determine their order or rank in the list. The adjective terms first, second, third, fourth, fifth, and so on are used to indicate the place of an object or person in the hierarchy. The order or sequence of the ordinal numbers changes depending on the criteria used to mark the positions, such as size, weight, markings, etc. Ordinals is another name for ordinal numbers. An example of ordinal numbers would be **Rose stood first in his class**. Here, first is the ordinal number describing the position of Rose in her class. Ordinal, Nominal, and Cardinal Numbers are the three types of numbers used to represent objects in mathematics.

We might begin with listing the choices or options that can be discrete or continuous.

Discrete: {Diet Coke, Diet Pepsi, diet root beer, 7UP}
{Diet Coke, Diet Pepsi, 7UP}
{Negotiate, Go to war, Acquiesce, Capitulate}
{arm, disarm}
US {do nothing blockade, air strike} USSR {withdraw missiles, maintain missiles}
Predator {attack, chase} Prey {run, hide}

Continuous: prices $\times \geq 0$, level of effort $0 \leq x \leq 100$, level of benefit, level of readiness, UA Army Physical Fitness test score $0 \leq x \leq 300$
First, we consider preferences

tastes, likes/dislikes

not whether inherently inferior or superior

normative theory: what should our preferences be?

positive theory: given preferences, what do they imply?

First, it is essential that a complete ordering must be done.
Given our choices {Diet Coke, Diet Pepsi, diet root beer, 7UP}
This ordering might be:

Diet Coke > diet root beer >Diet Pepsi > 7UP.

If I like Diet Coke better than diet root beer and I like diet root beer better than Diet Pepsi. Then I can infer that I like Diet Coke better than Diet Pepsi.
We cannot violate our own logic. Where do we start?
Ordinal rankings or ordinal utility:

Diet Coke – like best

Diet Root Beer – like next best

Diet Pepsi – like next least

7UP – like least

So we could assign points to first, second, third, fourth place with points such as: 4, 3, 2, 1.

The point values, payoffs, signify only that we like one better than another.

We point out that since these numbers are ordinal we cannot do REAL math with these numbers. We cannot do subtraction, addition, ratios, etc. They just don't make sense to do mathematics.

7.3 Cardinal Numbers

DEFINITION

Cardinal numbers are numbers that are used for counting real objects or counting things. They are also known as "counting numbers" or "cardinals". We commonly use cardinal numbers or cardinals to answer the question starting with "How many?"

So, why are they important in mixed strategies? Cardinal numbers are real numbers where the value has meaning: weight, height, age, income, number of casualties, etc.

It makes sense that we can do mathematics with cardinal numbers. If we have four people and they weigh {205, 193, 184, 160}, then our heaviest person weighs 12 lbs over the next highest weight.

In Cardinal Utility, we might use expected value points; we might use real data {time, counts, etc.}, or we might not have either available so now what?

7.4 Utility

Utilities are used to measure preferences. A preference is what a player prefers. For example, if Rose prefers A over B then A has a higher cardinal number than B.

Example 7.1: 3 × 2 Game

		Colin	
		C1	**C2**
Rose	R1	u	v
	R2	w	x
	R3	y	z

Perhaps the best Rose can do here is to provide an ordering such as I prefer the results or R_iC_j as u, w, x, z, y, v.

We must also have preferences for Colin. In a zero-sum game, Colin's preferences must be the opposite of Rose's preferences. Thus, Colin's preferences must be v, y, z, x, w, u.

We might assign numbers to those who follow these preference rules.

Example 7.2: 3 × 2 Game with Values

		Colin	
		C1	C2
Rose	R1	100	50
	R2	90	80
	R3	60	70

First, if we solved this game by the methods in Chapter 5, we would find a saddle point (pure strategy) solution at R2C2 with payoffs (80, −80). As a matter of fact, any 3 × 2 game with entries in this numerical order would result in a saddle point solution at R2C2.

Although ordinal numbers are adequate for pure strategy solutions, we recommend always converting to cardinal utilities.

Example 7.3: 2 × 2 Game

		Colin	
		C1	C2
Rose	R1	a	b
	R2	c	d

Let's assume Rose likes a better b so $a > b$ and Rose likes d better than c. In this case, we would not have a saddle point solution and would require to find a mixed strategy solution. Since the values (by oddments) use a − b and d − c with $(a - b) + (c - d)$ as the denominator then the actual values of a, b, c, and d do make a difference.

The game theory solution would be

$$R_1C_1 \frac{(d-c)}{(a-b)+(d-c)}, R_2C_1 \frac{(a-b)}{(a-b)+(d-c)}$$

If $a = 6, b = 3, d = 4, c = 2$ then the mixed strategy probabilities would be 2/5 for R1C1 and 3/5 for R2C1.

If $a = 10, b = 3, d = 11, c = 3$ then the mixed strategies probabilities would be 8/15 for R1C1 and 7/15 R2C1.

These certainly do not provide the same answer to the game for the strategies to play as probabilities.

Von Neumann and Morgenstern created a way to do this: called *lottery*. It is more time-consuming and tedious especially if there are lots of choices. Assume we have only four choices: u, x, w, v. We might rank our preferences as *u, x, w*, and *v*.

Here is basically how it works:

Pick a point spread (100 best, 0 worse) chosen arbitrarily
Your best Diet Coke, *u*, can be scored as worth 100 points
The lowest 7UP, *v*, might be scored as worth 0 point
What are diet root beer, *x*, and Diet Pepsi, *w*, worth?

So next we ask Rose something like, "which would you prefer: x for certain or a lottery that gives you ½ of u and ½ of v". We denote this lottery as ½ *u* ½ *v*. Now, ½ *u* ½ *v* = 50. If Rose likes *x* more than 50 points, then we repeat the lottery with maybe ¼ *v* ¾ *u*, which is equal to 75. Let's assume Rose now prefers the lottery so we know our value of *x is between (50, 75)*. We continue this process until we get value for *x*. Let's assume we stop at 64 for *x*.

We move on to *w* and do a similar lottery process but our end points will be *v* and *x* now. We might start with ½ *v* ½ *x* = 32. Let's assume Rose takes this lottery so *w* = 32.

Our cardinal values will be 100, 64, 32, 0 for utility values for our preferences.

Example 7.4: Utilities

Consider a game with outcomes *u, v, w*, and *x*. We also have the preference ordering of *u, x, w, v*. To assign cardinal scale numbers, we could ask and answer questions about lotteries. Assign numbers to the end points arbitrarily so that the values are *u* > *v*. For example, let *u* = 100, *v* = 0. Now ask "would you prefer x for certain or a lottery that provides *u* with probability ½ and v with probability ½". If we prefer x to the lottery than x has a value greater than the midpoint (50): now ask "would you prefer x for certain or a lottery that provides *u* with probability 1/4 and *v* with probability 3/4". If this time we prefer the lottery more than x then x has a value between (50, 75). We continue to stop at a number.

Let's assume we gave DRB 62 points and 7UP 25 points.

Example 7.5: Vending Machines

Let's assume that we have two options today:

1. Go to vending machine #1 which is out of Diet Coke, so you go for second best and get DRB because that is all they have left.
2. Go to a machine #2 that gives DC that has p(Dc) = 0.5, p(DRB) = 0.20, P(7UP) = 0.25, and p(DP) = 0.05

Which is better?

$$E[\text{Machine 1}] = 62$$

$$E[\text{Machine 2}] = 0.5\ (100) + 0.2\ (62) + 0.25\ (25) + 0.05\ (1)$$
$$= 50 + 12.4 + 8.25 + 0.05 = 70.69$$

Machine 2 is better because it yields a higher value according to our preference scale.

Example 7.6: Game Payoff Matrix

Next issue: everyone might have a different scale. Does this matter? Let's see.

```
              Colin
              C1  C2
Rose  R1   w    x
      R2   y    z
```

Assume a zero-sum game.
Preference ordering $x > w > y > z$
Pick numbers for yourself
$x = 10, w = 8, y = 6, z = 4$
What happens?

```
              Colin
              C1  C2
Rose  R1  8   10
      R2  6    4
```

Regardless of the numbers used, the solution is R1C1 and the value placed in that position.

Example 7.7: Larger 3 × 2 Game Payoff Matrix

```
Rose    R1   u    v
        R2   w    x
        R3   y    z
```

Maybe the best Rose can do is give an ordering say *u, w, x, z, y, v* (*big is better*).
Maybe we go to Colin and he gives an ordering say *v, y, z, x, w, u* (*small is better*).

We can assign numbers to their preference.
Let $u = 10, w = 9, x = 8, z = 7, y = 6, v = 5$
I like u twice as much as v.

```
              Colin
              C1  C2
Rose  R1  10   5
      R2   9   8
      R3   6   7
```

Solution is always R2C2, your game value differs only because of preference.

Example 7.8: Now Consider, a Game Where a > b, and d > c

We will always use oddments

Rose		Colin C1	C2	Oddment
	R1	a	b	$d - c$
	R2	c	d	$a - b$

Rose		Colin C1	C2	Oddment
	R1	5	3	2; 6/8
	R2	2	8	6; 2/8
		3	5	
		5/8	3/8	

V = 34/8
But what if

Rose		Colin C1	C2	Oddment
	R1	10	1	9; 1/10
	R2	2	3	1; 9/10
		8	2	
		.2	.8	

V = 28/10 = 2.8

Both the oddments probability and game value change but we are still using oddments. For a mixed strategy total conflict game, that values make a difference.

Example 7.9: Consider the Following Game

```
    C1 C2
R1  R  S
R2  T  U
```

Consider the game

Rose's preference t to s to r to u. order is t > s > r > u
Rose is indifferent between s̲ and the lottery 2/3t, 1/3r.
Rose is indifferent between r and the lottery 1/2s,1/2u.
Rose's other preferences are consistent with these.
Colin's preferences are the opposite of Rose's

How should the game be played? If Rose is offered a choice of s or to play the game which should she choose?

$t > s > r > u$: Assume OR LET the lottery be $t = 100$, $u = 0$ for the best and worst case.

$$s = 2/3t + 1/3 r \rightarrow s = 2/3 (100) + 1/3r = 200/3 + 1/3 r$$

$$r = 1/2s + 1/2u \rightarrow r = 1/2s + 1/2 (0) = \frac{1}{2} s$$

Solve these two equations and two unknowns: $s = 1/3r + 200/3$ and $r = 1/2\, s$ and we find $s = 80$ and $r = 40$.

$100 > 80 > 40 > 0$ (works).

There is no dominance and no saddle point. Therefore, we use mixed strategies

Rose	Colin				
	C1	C2			
R1	40	80	40	100/140 = 5/7	
R2	100	0	100	40/140 = 2/7	V = 400/7
	60	80			
	8/14	6/14			
V = 400/7					

$400/7 = 57.14$ which is less than 80 we take s. We do not play the game.

Observations: if in a zero-sum game, the largest values are on a diagonal, then we will use oddments to solve. Otherwise, look for a saddle point.

7.5 Von Neumann-Morgenstern Utilities Applied to Game Theory

Suppose we have three alternatives R = {R1, R2, R3} and assume that you have a preference and like R1 > R2 > R3. Suppose we arbitrarily assign values to R1 and R3, the best and worst alternatives of 200 for R1 and 100 for R3. Now we need a value of R2.

Let R1: McDonald's Quarter Pounder Combo

Let R2: Burger King's Whopper Combo

Let R3: Wendy's Single Combo

Break into group of two:

Through questions attempt to fix the value of R2 between R1 and R3.

1. Do you prefer R2 with probability 1 or a lottery that gives you R1 with probability ½ and R3 with probability ½. Now, if you prefer

R2 to the lottery you know R2 ranks higher than the midpoint, 150. Otherwise, R2 falls between 100 and 150.

2. Do you prefer R2 with probability 1 or a lottery that gives you R1 with probability 3/4 and R3 with probability 1/4. Now, if you prefer the lottery you know R2 ranks higher than the midpoint, 150 but less than 175.

3. At some point, you are indifferent to the selection of R2 and the lottery, assume it is at R1 with 0.70 and R3 at 0.3. Thus, the value of R2 = 170.

```
        Colin
        C1  C2
Rose R1  w   x
     R2  y   z
```

Assume zero-sum game.
Preference ordering is $x > y > w > z$
$10 > 8 > 6 > 4$

```
        Colin
        C1  C2
Rose R1  6   10
     R2  8   4
```

In class, we discussed:

```
        Colin
        C1  C2
Rose R1  w   x
     R2  y   z
```

Assume zero-sum game.
Preference ordering $x > w > y > z$
Via picking values
If we use AHP with the following matrix:

		x 1	w 2	y 3	z 4
1	x	1	3	5	7
2	w	1/3	1	2	4
3	y	1/5	1/2	1	3
4	z	1/7	1/4	1/3	1

I get weights (eigenvector) of the following (to three decimals)

$X = 0.595$
$W = 0.211$
$Y = 0.122$
$Z = 0.071$

If we put these values into the payoff matrix then our solution is still R1C2 but the value of the game is now 0.595.

Thus, AHP or TOPSIS can help obtain the relative values of the outcomes.

7.6 An Alternative Approach to the Lottery Method in Utility Theory for Game Theory

7.6.1 Ordinal versus Cardinal Utility

Ordinal utility is a method that ranks outcomes. We tell our students it is like knowing the names of how people finish in a race, first, second, third, …, last. Cardinal utility uses interval scale values where we would now replace the order of finish with the *times* they ran the race. With the times, we know how much faster each runner is compared to the other runners.

Often real data are not available for analysis in a game theory scenario. Perhaps the best students can initially do is "rank order" the outcomes from 1 to n for each player in the game.

7.6.2 Lottery Method Illustrated

Consider an example where we have a choice between going to McDonald's or going to Burger King. Assume that we limit ourselves to the following meal choices:

1. Burger King: Whopper and Fries Combo (x), Whopper Jr and Fries Combo (y)
2. McDonald's: Big Mac and Fries Combo (w), Quarter pounder $ fries combo (z)

Step 1. We need an ordinal preference for these choices. Let's assume the row preferences are:

$$z > x > y > w$$

Step 2. Use the lottery method to assign values: start by assigning z and w arbitrarily keeping in mind that z gets a higher value than w. We could use a scale from (0, 100) and assign 100 to Z and 0 to W, as an example.

Step 3. Next, consider x. Would you prefer x for certain or a lottery which gives you z at 50% of the time and w at 50% of the time? ½ z ½ w. If Rose likes x over the lottery, then x ranks higher than the midpoint between z and w. So we use numbers greater than 50. So you try, would you prefer x for certain or a lottery that gives ¼ w ¾ z? Now, if Rose prefers the lottery, then x has a value between 50 and 75. We continue until we narrow the value to a point. When Rose is indifferent between the certainty and the lottery, we are done. Assume this occurs at 40% w and 60% z. We then would take 60% of 100 for the value of x.

Step 4. We do the same thing for y. Assume, we go through our process and assign a value of 20 for y.

Step 5. Now, become the column player.

Steps 6–9. Repeat Steps 1–4 to obtain values for the column player's preferences.

This could eventually lead to the following payoff matrix assuming the column player's preferences are directly at odds with the row player. The result would be a pure strategy solution where Player 1 gets his third choice and Player 2 gets his second choice, shown in Table 7.1.

7.6.3 AHP Method

AHP and AHP-TOPSIS hybrids have been used to rank order alternatives among numerous criteria in many areas of research in business industry and government (Fox, 2014a, 2014b) including such areas as social networks (Fox et al., 2013, 2014), dark networks (Fox et al., 2014), terrorist phase planning (Fox et al., 2014; Thompson and Fox, 2014), and terrorist targeting (Fox, 2014a, 2014b).

Table 7.2 represents the process to obtain the criteria weights using the analytic hierarchy process used to determine how to weight each criterion for

TABLE 7.1

Payoff Matrix for Lottery Example

		Player 2	
		C1	C2
Player 1	R1	(100, 0)	(40, 80)
	R2	(60, 20)	(0, 100)

TABLE 7.2

Saaty's Nine-Point Scale

Intensity of Importance in Pairwise Comparisons	Definition
1	Equal importance
3	Moderate importance
5	Strong importance
7	Very strong importance
9	Extreme importance
2, 4, 6, 8	For comparing between the above
Reciprocals of above	In comparison of elements i and j, if i is 3 compared to j, then j is $1/3$ compared to i.
Rationale	Force consistency; measure values available

the TOPSIS analysis. Using Saaty's nine-point reference scale (Saaty, 1980), displayed in Table 7.2, we used subjective judgment to weight each criterion against all other criteria lower in importance. Figure 7.1 displays the template used.

Now, assume we have a game where we might know preferences in an ordinal scale only.

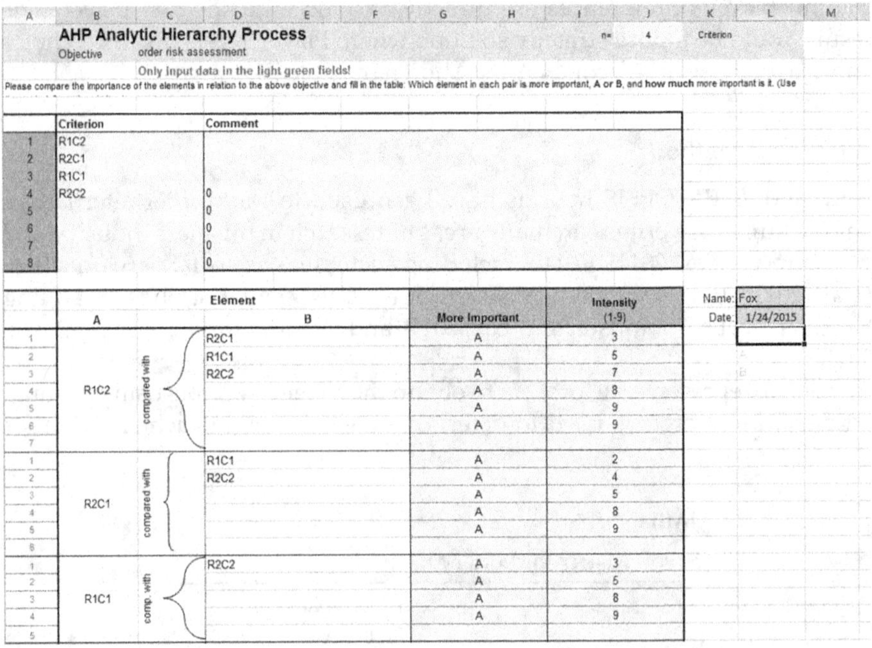

FIGURE 7.1
AHP template.

```
           Player 2
           C1  C2
Player 1 R1  w   x
         R2  y   z
```

Also, assume that this is a zero-sum game.

Player 1's preference ordering is $x > y > w > z$. Now we might just pick values that meet that ordering scheme, such as $10 > 8 > 6 > 4$ yielding

```
           Player 2
           C1  C2
Player 1 R1  6   10
         R2  8    4
```

We obtained the following AHP matrix:

		x	w	y	z
		1	2	3	4
1	x	1	3	5	7
2	w	1/3	1	2	4
3	y	1/5	1/2	1	3
4	z	1/7	1/4	1/3	1

We get weights (eigenvector) of the following (to three decimals)

$x = 0.595$

$w = 0.211$

$y = 0.122$

$z = 0.071$

```
             Player 2
             C1      C2
Player 1 R1  0.211   0.595
         R2  0.122   0.071
```

The solution, regardless of the number put in for w, x, y, or z, is the value in R1C2. The difference is the method using AHP is based on preferences. Thus, AHP can help obtain the relative values of the outcomes. These values are the cardinal utilities values based upon the preferences.

7.6.4 AHP Example in Game Theory

In our game theory course, we initially cover ordinal utility as a method to obtain values for a payoff matrix. Let's apply this to two-person non-zero-sum game example from the course.

Example 7.10: Unites States versus Country X

Consider a game between two players with two strategies each where the best we can initially do is to ordinally rank their preferences. The game payoff matrix is listed in Table 7.3.

TABLE 7.3

Ordinal Payoff Matrix

		Country X	
		C1	C2
United States	R1	(2, 4)	(4, 1)
	R2	(3, 2)	(1, 3)

There are no pure strategies so the players must play equalizing or mixed strategies to find an equilibrium. We find that we are stuck because these are ordinal values. In the past, our students just assumed that these values are in fact cardinal values. With that assumption, we find the United States plays ¼ R1 and ¾ R2, while Country X plays ¾ C1 and ¼ C2. The Nash equilibrium is (2.5, 2.5). Further, if we find Prudential Strategies, the Security Values, to go to Nash Arbitration with these values, we find that the United States plays ½ R1, ½ R2 with a security value of 2.5, while Country X plays ½ C1, ½ C2 with a security value of 2.5. Using (2.5, 2.5), we find that the Nash Arbitration values are (2.75, 2.875) while playing 3/8 of R1C2 and 5/8 of R1C1, as displayed in Figure 7.2 using the AHP method (Saaty, 1980).

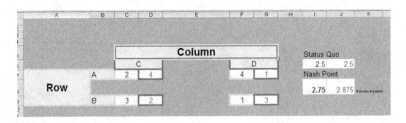

FIGURE 7.2
Screenshot.

The issue is "what does that mean" since the initial values were merely ordinal values with no indication of how much better a 4 is than a 3, 2, or 1 for each player.

Rather than using the Lottery Method suggested by Morgenstern and von Neumann, we suggest the pairwise comparison method of Saaty for each player's strategy combinations. For both the United States and Country X, we will need cardinal values for their preferences with these combined strategies: *R1C1*, *R1C2*, *R2C1*, and *R2C2*.

First, we use Saaty's method (Saaty, 1980) for the United States. We utilize a template build for classwork (Fox, 2012), Figure 7.3.

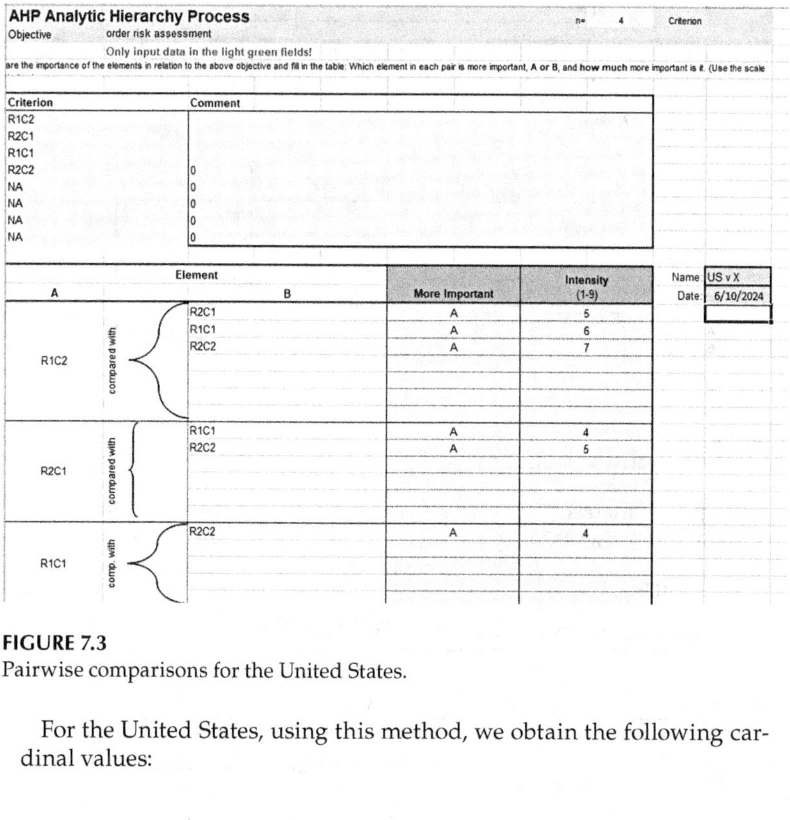

FIGURE 7.3
Pairwise comparisons for the United States.

For the United States, using this method, we obtain the following cardinal values:

R1C2	0.649830851
R2C1	0.17166587
R1C1	0.105026015
R2C2	0.073477264

For Country X, we obtain cardinal values as shown in Figure 7.4. The cardinal values for Country X are as follows:

R1C1	0.612431976
R2C2	0.243316715
R2C1	0.091240476
R1C2	0.053010833

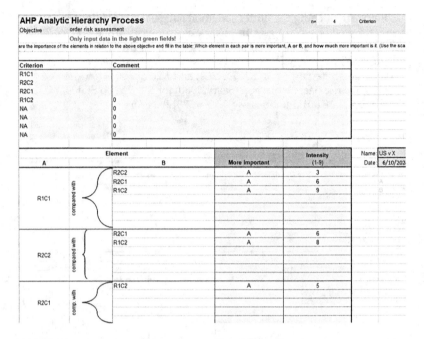

FIGURE 7.4
Pairwise comparisons for Country X.

The payoff matrix, with cardinal values representing true preferences, is displayed in Table 7.4.

TABLE 7.4

Cardinal Payoff Matrix Using AHP Results

		Country X	
		C1	C2
United States	R1	(0.1050, 0.6124)	(0.6498, 0.0530)
	R2	(0.1717, 0.0912)	(0.0735, 0.2433)

The Nash equilibrium, Prudential Strategies, and the Nash Arbitration are found using templates built for classroom use (Feix, 2007). We find the Nash equilibrium (0.202619, 0.16153).

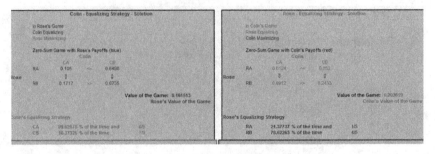

We find the Prudential Strategies or Security Levels are the Nash equilibrium from before.

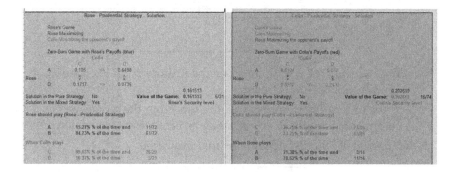

We find the Nash Arbitration (0.373, 0.3368) by playing 0.5075 of R1C1 and 0.4925 of R1C2.

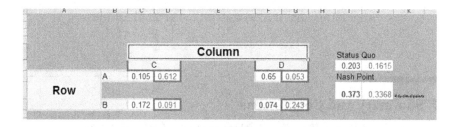

We see that our mixed strategies probabilities are different with cardinal preferences than they were with the ordinal preferences that we merely assumed were cardinal preferences. We have had cases where the decisions in TOPSIS and game theory are altered through the use of this method to obtain cardinal values as well as sensitivity analysis of the cardinal weights.

Chapter 7 Exercises

7.1 Assume the utilities for u, x, v, w are 100, 60, 30, 10, respectively, state which alternative Rose prefers in each pairs when

a. x versus ¾ w, ¼ u

b. x versus ½ w, ½ u

c. ¼ w ¾ x versus ½ u, ½ v

7.2 Suppose we have the following zero-sum game payoff matrix:

		Colin	
		C1	C2
Rose	R1	u	v
	R2	w	x

Let the values for u, v, w, and x be the utilities from Exercise 7.1
a. Find Rose's optimal strategy
b. If the order of preferences was the same, but the values were different, what would Rose's strategy be under these conditions

7.3 If a partial conflict game with ordinal utilities is:

		Colin	
		C1	C2
Rose	R1	(1, 4)	(4, 2)
	R2	(2, 2)	(3, 3)

Use either lottery or AHP to convert to cardinal utilities.

7.7 Summary and Conclusions

We have shown that differences in playing strategies in game theory occur as a function of the values in the payoff matrix. Table 7.5 displays a comparative summary of our example.

In conclusion, not only did the numerical values change but two key points are seen in this example. First, using cardinal values, the Nash arbitration favors the United States where as before it favored Country X. Second, how we play our strategies in the game changes substantially.

TABLE 7.5

Summary Results

Results	Ordinal Values	Strategies Played	Cardinal Values	Strategies Played
Nash Equilibrium	(2.5, 2.5)	¼ R1, ¾ R2, ¾ C1, 1/4 C2	(0.20219, 0.161513)	1/5 R1, 4/5 R2, 8/9 C1, 1/9 C2
Security Level	(2.5, 2.5)	½ R1, ½ R2, ½ C1, ½ C2	(0.20219, 0.161513)	11/72 R1, 61/72 R2, 26/29 C1, 3/29 C2
Nash Arbitration	(2.75, 2.875)	3/8 R1C2, 5/8 R1C1	(0.373, 0.3368)	0.5075 R1C1, 0.4925 R1C2

References

Fox, W. (2014a). Chapter 221, TOPSIS in business analytics. *Encyclopedia of Business Analytics and Optimization.* Levi and Garcia, Eds. IGI Global and Sage Publications. V (5), pp. 281–291.

Fox, W.P. (2012). Mathematical modeling of the analytical hierarchy process using discrete dynamical systems in decision analysis. *Computers in Education Journal,* 3 (3), pp. 27–34.

Fox, W.P. (2014b). Using multi-attribute decision methods in mathematical modeling to produce an order of merit list of high valued terrorists. *American Journal of Operation Research,* 4 (6), pp. 365–374.

Saaty, T. (1980). *The Analytical Hierarchy Process.* United States: McGraw Hill.

Thompson, N. and W. Fox. (2014).Phase targeting of terrorist attacks: Simplifying complexity with TOPSIS. *Journal of Defense Management,* 4 (1), pp. 1–6. http://dx.doi.org/10.4172/2167-0374.1000116.

Additional Readings

Feix, M. (2007). Game theory: Toolkit and workbook for defense analysis students. M.S. Thesis, Department of Defense Analysism Naval Postgraduate School, Monterey, CA.

Fox, W. & M.N. Thompson. (2014). Phase targeting of terrorist attacks: Simplifying complexity with analytical hierarchy process. *International Journal of Decision Sciences,* 5 (1), pp. 57–64.

Fox, W. & S. Everton. (2013). Mathematical modeling in social network analysis: Using TOPSIS to find node influences in a social network. *Journal of Mathematics and Systems Science,* 3 (2013), pp. 531–541.

Fox, W. & S. Everton. (2014). Mathematical modeling in social network analysis: Using data envelopment analysis and analytical hierarchy process to find node influences in a social network. *Journal of Defense Modeling and Simulation.* 2014, pp. 1–9. (published online, journal due in summer 2014)

Fox, W. & S.F. Everton. (2014). Using mathematical models in decision making methodologies to find key nodes in the Noordin Dark Network. *American Journal of Operations Research,* (July), pp. 1–13 (online).

Straffin, P. (2004). *Game Theory and Strategy.* Washington, DC: The Mathematical Association of America: New Mathematics Library.

8

Nash Equilibrium and Non-Cooperative Solutions in Partial Conflict Games

---

8.1 Introduction

In a two-person game where the payoff entries do not sum to zero nor to the same constant, then we have a partial conflict game (PCG). Further, if players do not communicate nor cooperate then similar methods to total conflict games might be used to find the Nash equilibrium. In total conflict games, the Nash equilibrium is the solution to the game. This may not be true in partial conflict games where we might consider additional strategies in finding a solution. Let's begin by defining a Nash equilibrium again.

NASH EQUILIBRIUM DEFINITION

The **Nash** equilibrium is a decision-making theorem within game theory that states a player can achieve the desired outcome by not deviating from their initial strategy.

In the Nash equilibrium, each player's strategy is optimal when considering the decisions of other players. Every player wins because everyone gets the outcome they desire.

Theorem: If players can achieve better payoff by unilaterally changing their strategies, then we have a Nash equilibrium.

Even if we find a pure strategy equilibrium by the movement diagram, we must also look for the equilibrium or equilibriums by other methods. Additionally, even if a partial conflict game has a pure strategy solution it might also have an equalizing strategy solution. The methods we present are: equalizing strategies by hand, calculus, and nonlinear optimization.

8.2 Pure Strategies and Dominance Review in Symmetric Games

We may use movement diagram to look for pure strategy solution in PCG.
In a movement diagram, for Rose:
Draw an arrow from the smaller to the larger values in the Rows
In a movement diagram, for Colin:
Draw an arrow from the smaller to the larger values in the Columns
If all the arrows point to one or sets of strategies, then we have pure strategy solutions.

Example 8.1: Pure Strategy Game

		Colin	
		C1	C2
Rose	R1	(2, 3)	(3, 2)
	R2	(1, 0)	(0, 1)

Arrows show R1C1 (2, 3) is the Nash equilibrium.

Example 8.2: Pure Strategy Two Solutions

		Colin	
		C1	C2
Rose	R1	(1, 1)	(2, 5)
	R2	(5, 2)	(−1, −1)

Here we have the arrows giving R2C1 at (5, 2) and R1C2 at (2, 5) as Nash equilibriums. Here we point that neither is achievable.

Example 8.3: Pure Strategy Solution

		Colin	
		C1	C2
Rose	R1	(3, 3)	(−1, 5)
	R2	(5, −1)	(0, 0)

Arrows point to (0, 0) at R2C2 as the Nash equilibrium. We clearly see a better solution for both players at R1C1 at (3, 3).

Example 8.4: Pure Strategy Two Solutions

		Colin	
		C1	**C2**
Rose	R1	(0, 0)	(2, 1)
	R2	(1, 2)	(1, 1)

We have two unachievable equilibriums at R2C1 and R1C2.

Example 8.5: No Pure Strategy Solution

		Colin	
		C1	**C2**
Rose	R1	(2, 4)	(1, 0)
	R2	(3, 1)	(0, 4)

Here the arrows do not point to one or more sets of strategies. When this happens, we must use equalizing strategies to find the Nash equilibrium.

There is no pure strategy equilibrium. We move counterclockwise around the strategies.

8.3 Equalizing Strategies

Therefore, we need to find the Nash equilibrium by equalizing strategies. Equalizing strategies are used to find the mixed strategy equilibrium (only use if no pure strategy). We decided to start with Rose's equalizing strategy. After that we next equalize Colin's expected value by finding Rose's mixed strategy in Colin's game with Colin's equalizing strategy.

We equalize Rose's expected value with finding Colin's mixed strategy in Rose's game.

	Colin	
	C	**D**
A	(2, 4)	(1, 0)
B	(3, 1)	(0, 4)

Rose's game

	Colin	
	C	D
A	2	1
B	3	0
	1/2	1/2
Vr = 3/2		

Colin's game

	Colin		
	C	D	
A	4	0	3/7
B	1	4	4/7
Vc = 16/7			

We find (3/2, 16/7) is the Nash equilibrium by equalizing strategies. We might use calculus to find these values.

8.3.1 Calculus

Create two functions: Let A_{nxm} be the payoff to Player 1 and B_{nxm} be the payoff to Player 2.

$$E_1(x, y) \ \& \ E_2(x, y), \text{ where } E1(X,Y) = XAY^T \text{and } E2(X,Y) = XBY^T$$

Take the partial derivatives and solve the system equal to 0.

$$\frac{\partial E_1}{\partial x_i} = 0 \text{ and } \frac{\partial E_2}{\partial y_j} = 0 \text{ for } i = 1, 2, \ldots, n-1 \text{ and } j = 1, 2, \ldots, m-1$$

If there is a solution that satisfies the constraints, $x_i \geq 0$, $y_j \geq 0$ and $\Sigma\, x_i \leq 1$ and $\Sigma\, y_j \leq 1$, then this is a mixed strategy Nash equilibrium.

Example 8.6: Repeat Example 8.4 using Calculus Methods

	Colin	
	C	D
A	(2, 4)	(1, 0)
B	(3, 1)	(0, 4)

$$A = \begin{bmatrix} 2 & 1 \\ 3 & 0 \end{bmatrix}, B = \begin{bmatrix} 4 & 0 \\ 1 & 4 \end{bmatrix}$$

$$E[XY] = (-x+3)y + x(1-y)$$

$$\frac{\partial E}{\partial x} = -y + (1-y)$$

We set

$$\frac{\partial E}{\partial x} = -y + (1-y) = 0$$

We find y = ½ and 1 − y = ½.

$$\frac{\partial E}{\partial y} = 4x + 3x - 4$$

$$\frac{\partial E}{\partial y} = 7x - 4 = 0$$

$$x = \frac{4}{7} \text{ and } 1 - x = 3/7$$

We substitute back into E[XY] to get (3/2, 16/7), our Nash equilibrium by equalizing strategies.

So, $y_1 = y_2 = 1/2$ and $x_1 = 4/7$ and $x_2 = 3/7$ as before with value of the game (16/7, 3/2).

We might also use nonlinear programming (NLP).

Method 3 NLP: NLP Approach for Two or More Strategies for each Player

For games with two players and more than two strategies each, we present the nonlinear optimization approach by Barron (2013). Consider a two-person game with a payoff matrix as before. Let's separate the payoff matrix into two matrices **M** and **N** for Players 1 and 2 We solve the following nonlinear optimization formulation in expanded form, in Equation (8.1).

$$\text{Maximiz} \sum_{i=1}^{n}\sum_{j=1}^{m} x_i a_{ij} y_j + \sum_{i=1}^{n}\sum_{j=1}^{m} x_i b_{ij} y_j + -p - q$$

Subject to:

$$\sum_{j=1}^{m} a_{ij} y_j \le p, i = 1, 2, \ldots, n,$$

$$\sum_{i=1}^{n} x_i b_{ij} \le q, j = 1, 2, \ldots, m, \tag{8.1}$$

$$\sum_{i=1}^{n} x_i = \sum_{j=1}^{m} y_j = 1$$

$x_i \geq 0, y_j \geq 0$

We return to our previous example. We define M and N as:

$$M = \begin{bmatrix} 2 & 1 \\ 3 & 0 \end{bmatrix} \text{ and } N = \begin{bmatrix} 4 & 0 \\ 1 & 4 \end{bmatrix}$$

We define x_1, x_2, y_1, y_2 as the probabilities for players playing their respective strategies.

By substitution and simplification, we obtain the NLP formulation. We illustrate EXCEL but many technologies are available to solve NLP problems.

Maximize $6y_1 x_1 + 4y_1 x_2 + x_1 y_2 + 4x_2 y_2 - p - q$

Subject to:

$x_1 + x_2 = 1$

$y_1 + y_2 = 1$

$4x_2 - q \leq 0$

$4x_1 + x_2 - q \leq 0$

$2y_1 + y_2 - p \leq 0$

$3y_1 - p \leq 0$

Non-negativity

	A		
		2	1
		3	0
	B	4	0
		1	4
dv			
x1		0.428572	
x2		0.571429	
p		1.500001	
y1		0.5	
y2		0.5	
q		2.285715	

We find the exact same solution as before, shown in the larger screenshot.

It is essential to note that in some partial conflict games there are both pure and equalizing Nash equilibriums.

Example 8.7: Games with both Pure and Mixed Strategies

		Colin C1	Colin C2
	R1	(2, 1)	(-1, 1)
Rose	R2	(-1, -1)	(1, 2)

The movement diagram yields two pure strategy equilibrium at R1C1 (2, 1) and R2C2 (1, 2). As players try to achieve these they end up at lower valued solutions. We next check for equalizing mixed strategy solutions. We find if Colin plays 0.4 C1 and 0.6 C2 with Rose playing 0.6 R1 and 0.4 R2 the solution is (0.4) (0.6) (2, 1) + (0.6) (0.6) (-1, -1) + (0.4) (0.4) (-1, -1) + (0.4) (0.6) (1,2) = (0.2, 0.2)

The Excel screenshot is shown in Figure 8.1

Finding a Solution

According to Straffin (2004), a Nash equilibrium is a solution if and only if it is unique and Pareto optimal. Pareto optimal refers to the northeast region of a payoff polygon where the payoff polygon is found as the convex set formed by the outcome coordinates, Figure 8.2.

We see in Figure 8.2 that the Nash equilibrium (1.5, 2.28) is not Pareto optimal and not the solution that we seek.

At this point, we might try to allow communication and try strategic moves which we do not describe here but can be reviewed in Giordano et al. (2013). Further, we might want to show the method of Nash arbitration although we do not illustrate that here.

We see we obtain the same results using either NLP or quadratic programming methods.

But is this Nash equilibrium the solution?

Payoff Polygon

A payoff polygon is a convex set of all the payoff coordinates. A convex set is defined as if we connect any two points in a set all the points on that line segment are in the set. See an illustration in Figure 8.3.

We will draw payoff polygon shown in Figure 8.4.

Pareto Principle: To be acceptable as a solution to a partial conflict game an outcome must be Pareto optimal.

Now, let's define **Pareto optimal**.

	A	B	C	D	E	F	G
	Decision Variables						
	x1	0.428571				Objective Function	
	x2	0.571429				z	0
	x3	0					
	x4	0					
	x5	0			X^TAY^T	1.5	
	x6	0			X^TBY^T	2.285714	
	x7	0					
	x8	0					
	x9	0					
	x10	0					
	P	1.5					
	v2	0					
	y1	0.5					
	y2	0.5					
	y3	0					
	y4	0					
	y5	0					
	y6	0					
	y7	0					
	y8	0					
	y9	0					
	y10	0					
	q	2.285714					
	v4	0					

FIGURE 8.1
EXCEL screenshot.

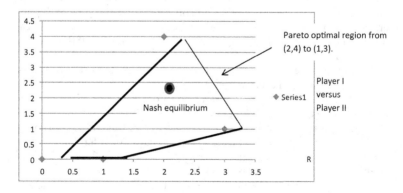

FIGURE 8.2
Payoff polygon and Pareto optimal region.

Convex   **Non-convex**

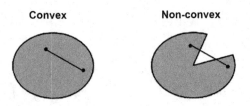

FIGURE 8.3
Convex and non-convex sets.

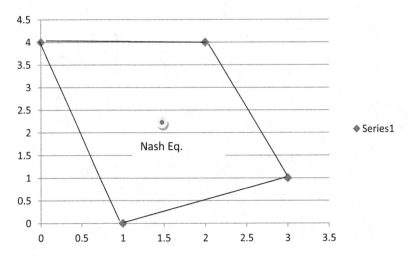

FIGURE 8.4
Payoff polygon for Example 8.1.

Pareto Optimal: An outcome is non-Pareto optimal (also known as Pareto inferior) if there is another outcome which would give both players higher payoffs or would give one player the same payoff and the other player a higher payoff. An outcome is Pareto optimal if there are no such other outcomes.

Pareto Principle: The Nash equilibrium does not lie on the northeast quadrant so it is not Pareto optimal.

The equalizing strategies are not Pareto optimal. The Nash equilibrium is inside the polygon and not on the NE side.

Perhaps, at this point in our analysis, we should introduce two more sets of strategy approaches: prudential and counter-prudential

Finding the prudential strategies for each player and then finding the counter-prudential strategies for each player.

Prudential Strategies: What is Rose's best strategy in Rose's game and What is Colin's best strategy in Colin's game? Thus, in prudential strategies, the players are trying to optimize their outcomes.

1. Extract the two game matrices: check for saddles points then mixed strategies.

Rose's Game   Colin's Game
2 1 x_1 4 0
3 0 x_2 1 4
                    y_1 y_2

Rose's best strategy is a saddle point (pure strategy) to always play Strategy A. By doing so, Rose is always guaranteed an outcome of at least 1. The value, 1, is Rose's security level and playing Strategy A is Rose's prudential strategy. Colin's best strategy is a mixed strategy in Colin's game playing 4/7 y_1 and 3/7 y_2. By playing this, Colin is guaranteed a value of at least 16/7. The value 16/7 is Colin's security level and Colin's prudential strategy is 4/7 y_1 *and* 3/7 y_2.

We note that the security level for the two players' point is (1, 16/7).

2. Counter-Prudential Strategies: These are a players' best strategy in opposition to the player's prudential strategies. Definition: In a non-zero-sum game, a player's <u>Counter-Prudential Strategy</u> is his optimal response to his opponent's prudential strategy.

If Rose thinks Colin is playing his prudential mixed strategy then she can find her "best" expected payoff:

Rose R_1 A: (4/7)*2 + (3/7)(1) = 11/7
Rose R_2 B: (4/7)*3 + (3/7) *0 = 12/7: <u>The better response since 12/7 ≥ 11/7.</u>

Thus Rose's counter-prudential strategy is always to play B.

If Colin thinks Rose is playing a prudential strategy of A, then Colin is better off always playing C.

Counter-prudential strategies summary Rose plays B and Colin plays C. Both are pure strategy decisions.

A summary is given in Table 8.1

Table. 8.1 Mix-Match Table of Output

Rose Strategy	Colin Strategy	Rose Payoff	Colin Payoff
Prud.	Prud.	1.57	2.29
Prud.	Counter PS	2	4
Counter PS	Prud.	1.71	2.29
Counter PS	Counter PS	3	1

If we plot these, we find none are Pareto optimal.

Nash Equilibriums and Formulations

Given a non-zero-sum game of two players with multiple strategies. Rose has payoffs in rows identified as *i* from 1 through *m*, and Colin has payoffs in Columns *j* from 1 through *n* where each pair is identified as

(M_{ij}, N_{ij}). Again, let's define the following payoff matrix that has components for both Rose and Colin:

$$(M,N) = \begin{bmatrix} (M_{1,1},N_{1,1}) & (M_{1,2},N_{1,2}) & \cdots & (M_{1,n},N_{1,n}) \\ (M_{2,1},N_{2,1}) & (M_{2,2},N_{2,2}) & \cdots & (M_{2,n},N_{2,n}) \\ \cdot & \cdot & \cdots & \cdot \\ \cdot & \cdot & \cdots & \cdot \\ \cdot & \cdot & \cdots & \cdot \\ (M_{m,1},N_{m,1}) & (M_{m,2},N_{m,2}) & \cdots & (M_{m,n},N_{m,n}) \end{bmatrix}$$

In non-cooperative non-zero-sum games, we use similar concepts of pure strategies and equalizing strategies (mixed strategy) to solve the game. We look for pure strategy solution using the *movement diagram*.

For Rose, she would maximize her payoffs, so she would prefer the highest payoff in each column. Arrows are in columns but values are Rose's. Similarly, Colin wants to maximize his payoffs, so he would prefer the high payoff at each row. We draw an arrow to the highest payoff in that row. Arrows are in rows but values are Colin's. If all arrows point in from every direction, then that or those points will be pure Nash equilibrium.

If all the arrows do not point at a value or values then we must use equalizing strategies to find the weights (probabilities) for each player. Basically, we would proceed as follows:

- Rose's Game: Rose Maximizing, Colin "Equalizing" is a zero-sum game which yields Colin's equalizing strategy.
- Colin's Game: Colin Maximizing, Rose "Equalizing" is a zero-sum game which yields Rose's equalizing strategy.
- Note: If either side plays their "equalizing strategy", the other side "unilaterally" cannot improve their own situation (stymie the other player).

This translates into two additional linear programming (LP) formulations, one for each maximizing player. We create two formulations that we call Equations (8.2) and (8.3) that when used provide the values of the game and the probabilities that the players should play their strategies in the equalizing strategy concept above.

$$\text{Maximize V} \tag{8.2}$$

Subject to:
$$N_{1,1}x_1 + N_{2,1}x_2 + \cdots + N_{m,1}x_n - V \geq 0$$
$$N_{2,1}x_1 + N_{2,2}x_2 + \cdots + N_{m,2}x_n - V \geq 0$$
$$\cdots$$
$$N_{m,1}x_1 + N_{m,2}x_2 + \cdots + N_{m,n}x_n - V \geq 0$$
$$x_1 + x_2 + \cdots + x_n = 1$$
$$x_i \leq 1 \quad for \quad i = 1,...,m$$
Nonnegativity

where the weights x_i yield Rose strategy and the value of V is the value of the game to Colin.

$$\text{Maximize } v \tag{8.3}$$

Subject to:

$$M_{1,1}y_1 + M_{1,2}y_2 + \cdots + M_{1,n}y_n - v \geq 0$$
$$M_{2,1}y_1 + M_{2,2}y_2 + \cdots + M_{2,n}y_n - v \geq 0$$
...
$$M_{m,1}y_1 + M_{m,2}y_2 + \cdots + M_{m,n}y_n - v \geq 0$$
$$y_1 + y_2 + \cdots + y_n = 1$$
$$y_j \leq 1 \quad for \quad j = 1,...,n$$
Nonnegativity

When the game is a 2×2 game, substitute the following equalizing rule for $x_i \leq 1$ and $y_j \leq 1$ constraints:

$$(N_{11} - N_{21})x_1 + (N_{12} - N_{22})x_2 = 0$$

and

$$(M_{11} - M_{21})x_1 + (N_{12} - N_{22})x_2 = 0$$

We will illustrate three examples.

Example 8.8 Equalizing Strategy Solutions

	Colin	
	C1	C2
Rose R1	(2, 4)	(1, 0)
R2	(3, 1)	(0, 4)

Formulation 1: Value to Colin equalizing by Rose so Colin's expected value of C1 equals his expected value of C2.

Max V

Subject to:

$$4x1 + 1x2 - V \geq 0$$
$$0x1 + 4x2 - V \geq 0$$
$$x1 + x2 = 1 \ \left(set \ the \ sum \ of \ the \ probabilities \ equal \ to \ 1\right)$$
$$(4 - 01) \ x1 + (1 - 4)x2 = 0 \ rewrite \ as \ 4x1 - 3 \ x2 = 0$$
Non-negativity

Formulation 2: Value to Rose equalizing by Colin so Rose's expected values $E[R1] = E[R2]$.

Max v

Subject to:
$$2y1 + y2 - v > 0$$
$$3y1 + 0y2 - v > 0$$
$$y1 + y2 = 1$$
$$(2-3)y1 + (1-0)y2 = 0 \text{ rewritten as } -y1 + y2 = 0$$
Non-negativity

Solve on EXCEL

Answers are in decimal form:

$$x1 = 0.428571 \quad x2 = 0.571429 \quad V = 2.285714$$

and

$$y1 = .5, \ y2 = .5, \ v = 1.5$$

These are the same as we found previously. The Nash equilibrium is (1.5, 2.285714).

Example 8.9: Larger than 2 × 2 Games

First, a strategy for 2 × n or m × 2 games using LP. Consider the following 4 × 2 game.

		Colin	
		C1	**C2**
Rose	R1	(−1, 3)	(5, −2)
	R2	(2, 1)	(4, 5)
	R3	(4, −2)	(−3, 6)
	R4	(5, 10)	(−4, −4)

Start with the game with the most strategies. In this case, Rose has four strategies. We will take Colin's game and apply the probabilities for Rose's four strategies. Why do we do this? We do it in hopes of reducing the game if any of x1, x2, x3, or x4 = 0 then we can remove those entries before solving what Colin should do in Rose's remaining game.

Max V

Subject to:
$$3x1 + x2 - 2x3 + 10x4 - V >= 0$$
$$-2x1 + 5x2 + 6x3 - 4x4 - V >= 0$$
$$x1 + x2 + x3 + x4 = 1 \ (sum \ of \ prob. = 1)$$
$$5x1 - 3x2 - 8x3 + 14x4 = 0 \qquad (equalizing \ V)$$
Non-negative

Solve on EXCEL
Solution to this game: $x1=0, x2=7/9, x3=0, x4=2/9$ $Vc=3$
Because $x1$ and $x3 = 0$ we can remove R1 and R3.

		Colin	
		C1	**C2**
Rose	R1	$(-1,3)$	$(5,-2)$
	R2	$(2,1)$	$(4,5)$
	R3	$(4,-2)$	$(-3,6)$
	R4	$(5,10)$	$(-4,-4)$

We formulate the second LP from the remaining entries R2C1, R2C2, R4C1, and R4C2.

We remove Row 1 and Row 3 since they will not be played, they have 0 probability.

Max v

Subject to:

$2y1 + 4y2 - v >= 0$

$5y1 - 4y2 - v >= 0$

$y1 + y2 = 1$

$-3y1 + 8y2 = 0$ (equalizing the two v's)

Non-negativity

Solve on EXCEL
Solution is $y1=8/11, y2=3/11$, $Vr=2.54=28/11$

Example 8.10 3 by 3 Game where Rose has 3 Strategies and Colin has 3 Strategies

The 3×3 game example and larger. There are no shortcuts other than if $m \neq n$ then solve the larger number of constraints first to see if we can eliminate any rows or columns.

Consider the non-zero-sum game with more than two strategies per player where there is no pure strategy equilibrium.

	Colin		
	c1	c2	c3
Rose			
r1	(9, 1)	(2, 2)	(7, 2)
r2	(3, 2)	(6, 1)	(4, 2)
r3	(5, 2)	(3, 2)	(5, 0)

Max Vc

Subject to:

$x1 + 2x2 + 2x3 - Vc \geq 0$

$2x1 + x2 + 2x3 - Vc \geq 0$

$$2x1 + 2x2 - Vc \geq 0$$
$$x1 + x2 + x3 = 1$$
$$-x1 + x2 >= 0 \text{ or } x1 \leq 1$$
$$-x1 + 2x3 >= 0 \text{ or } x2 \leq 1$$
$$-x2 + 2x3 >= 0 \text{ or } x3 \leq 1$$
All variables non-negative

and

Max Vr

Subject to:
$$9y1 + 2y2 + 7y3 - Vr \geq 0$$
$$3y1 + 6y2 + 4y3 - Vr \geq 0$$
$$5y1 + 3y2 + 5y3 - Vr \geq 0$$
$$y1 + y2 + y3 = 1$$
$$6y1 - 4y2 + 3y3 >= 0 \ \& \ y1 \leq 1$$
$$4y1 - y2 + 2y3 >= 0 \ \& \ y2 \leq 1$$
$$-2y1 + 3y2 - y3 >= 0 \ \& \ y3 \leq 1$$
All variables are non-negative

Solve using EXCEL
Solving these two LP problems yields the following results:
Vr=4.5, Vc=1.6 when Rose plays 0.4 R1, 0.4 R2, 0.2 R3 and Colin plays 0.0 c1, 0.25 c2, 0.75 c3. Thus the Nash equilibrium is (4.5, 1.6) and the probabilities to play strategies are $x1 = 0.4$, $x2 = 0.4$, $x3 = 0.2$, $y1 = 0$, $y2 = 0.25$, and $y3 = 0.75$.

8.4 Prudential Strategies with LP

We have Rose in Rose's game and Colin in Colin's game. So the LP formulations are in Equations (8.4) and (8.5).

$$\text{Maximize } V \tag{8.4}$$

Subject to:
$$N_{1,1}y_1 + N_{1,2}y_2 + \cdots + N_{1,m}y_n - V \geq 0$$
$$N_{2,1}y_1 + N_{2,2}y_2 + \cdots + N_{2,m}y_n - V \geq 0$$
$$\cdots$$
$$N_{m,1}x_1 + N_{m,2}x_2 + \cdots + N_{m,n}x_n - V \geq 0$$
$$y_1 + y_2 + \cdots + y_n = 1$$
$$y_j \leq 1 \quad for \quad j = 1,...,n$$
Nonnegativity

where the weights y_i yield Colin's prudential strategy and the value of V is the security level for Colin.

$$\text{Maximize } v \qquad (8.5)$$

Subject to:

$$M_{1,1}x_1 + M_{2,1}x_2 + \cdots + M_{n,1}x_n - v \geq 0$$
$$M_{1,2}x_1 + M_{2,2}x_2 + \cdots + M_{n,2}x_n - v \geq 0$$

...

$$M_{1,m}y_1 + M_{2,m}y_2 + \cdots + M_{m,n}y_n - v \geq 0$$
$$x_1 + x_2 + \cdots + x_m = 1$$
$$x_i \leq 1 \quad \text{for } i = 1, \ldots, m$$
Nonnegativity

Let's do Example 8.1 for Prudential strategies.

Example 8.11 Finding the Prudential Strategies

		Colin	
		C1	C2
Rose	R1	(2,4)	(1,0)
	R2	(3,1)	(0,4)

Rose's security level. Since we do not have equalizing strategies, we MUST replace the equalizing constraint with the constraints for each probability less than or equal to 1.

Max V

Subject to:
$$2x1 + 3x2 - V \geq 0$$
$$x1 + 0x2 - V \geq 0$$
$$x1 + x2 = 1$$
$$x1 \leq 1$$
$$x2 \leq 1$$

Solve on LINDO

LP OPTIMUM FOUND AT STEP   2

OBJECTIVE FUNCTION VALUE

1)   1.000000

VARIABLE	VALUE	REDUCED COST
V	1.000000	0.000000
X1	1.000000	0.000000
X2	0.000000	1.000000

ROW	SLACK OR SURPLUS	DUAL PRICES
2)	1.000000	0.000000
3)	0.000000	−1.000000
4)	0.000000	1.000000
5)	0.000000	0.000000
6)	1.000000	0.000000

NO. ITERATIONS=   2

Colin's security level.

Max v

Subject to:
$$2y1 + y2 - v \geq 0$$
$$3y1 + 0y2 - v \geq 0$$
$$y1 + y2 = 1$$
$$y1 \leq 1$$
$$y2 \leq 1$$
Non-negative

LP OPTIMUM FOUND AT STEP   3

OBJECTIVE FUNCTION VALUE

1)   2.000000

VARIABLE	VALUE	REDUCED COST
V	2.000000	0.000000
Y1	1.000000	0.000000
Y2	0.000000	1.000000

ROW	SLACK OR SURPLUS	DUAL PRICES
2)	0.000000	−1.000000
3)	1.000000	0.000000
4)	0.000000	2.000000
5)	0.000000	0.000000
6)	1.000000	0.000000

NO. ITERATIONS=   3

The prudential strategies are Rose plays R1 and Colin plays C1, and the security level is (2, 1).

Counter-prudential strategies cannot be solved via linear programming. They are fairly easy by hand.

If Rose plays R1, then Colin plays C1 to get 4.
If Colin plays C1, then Rose plays R2 to get 3.

8.5 Applications

8.5.1 Game Theory Applied to the Dark Money Network (DMN)

Discussing the strategies for defeating the DMN leads well into the game theory analysis of the strategies for the DMN and the State trying to defeat them. When conducting game theory analysis, we originally limited the analysis by using ordinal scaling and ranking each of the four strategic options 1–4. The game was set up below. Strategy R1 is for the state to pursue a non-kinetic strategy, and R2 is a kinetic strategy. Strategy C1 is for the DMN to maintain its organization, and C2 is to decentralize.

		Dark Money Network	
		C1	C2
State	R1	(2, 1)	(4, 2)
	R2	(1, 4)	(3, 3)

This ordinal scaling in the payoff matrix works when allowing communications and strategic moves; however, without a way of determining interval scaling, it was impossible to conduct analysis of prudential strategies, Nash arbitration, or Nash's equalizing strategies. We applied our method of analytical hierarchy process (AHP; Chapters 4 and 7) in order to determine the interval scaled payoffs of each strategy for both the DMN and the State. Again, we will use Saaty's standard nine-point preference in the pairwise comparison of combined strategies. For the State, the evaluation criteria we chose for the four possible outcomes were: how well it degraded the DMN, how well it maintained the state's own ability to fundraise, how well the strategy would rally its base, and finally, how well it removed nodes from the DMN. The evaluation criteria we chose for the DMN's four possible outcomes were: how anonymity was maintained, how much money the outcome would raise, and finally, how well the Koch brothers could maintain control of the network.

After conducting AHP analysis, we obtained a new payoff matrix with cardinal utility values.

		Dark Money Network	
		C1	C2
State	R1	(0.168, 0.089)	(0.366, 0.099)
	R2	(0.140, 0.462)	(0.324, 0.348)

With these cardinal scaling values, it is now possible to conduct a proper analysis that might include mixed strategies or arbitrated results (Chapter 10).

In this chapter, we will find just the equalizing and prudential strategies. Using a series of game theory solvers developed in EXCEL by Feix (2007), we obtain the following results.

- Nash Equilibrium: A pure strategy Nash equilibrium was found at R1C2 (0.366, 0.099) using strategies of non-kinetic and decentralize.
- Mixed Nash equalizing strategy is for the State to play non-kinetic 91.9% of the time and kinetic 8.1% of the time, and DMN to play maintain organization 100% of the time.
- Prudential strategies (security levels) (0.168, 0.099).

Since there is no equalizing strategy for the DMN, should the State attempt to equalize the DMN the result is as follows.

This is a significant departure from our original analysis prior to including the AHP pairwise comparisons in our analysis. The recommendations for the state were to use a kinetic strategy 50% of the time and a non-kinetic strategy 50% of the time. However, it is obvious that with proper scaling, the recommendation should have been to execute a non-kinetic strategy the vast majority (92%) of the time, and only occasionally (8%) conduct kinetic targeting of network nodes. This greatly reinforces the recommendation to execute a non-kinetic strategy to defeat the DMN.

8.5.2 Application through Illustrious Example: Writer's Guild Strike

In 2007–2008, the Writer's Guild of America went on strike. For more background information on the Writer's Guild Strike see Fox (2015b).

Let's assume the average TV DVD sells for $10.00. The request for an increase meant that the writers wanted to receive 0.6% instead of 0.3%. This amounts to an increase of $0.03 from $0.03 to $0.06 per sale.

8.5.2.1 Game Theory Approach

Let us begin by stating strategies for each player. Our players will be the Writer's Guild and the Management.

8.5.2.1.1 Strategies

Writer's Guild: Their strategies are to strike (S) or not to strike (NS).

Management: Salary increase plus revenue sharing as the writers requested (IN), revenue sharing from DVDs only at a rate of 0.6% (GIN), revenue sharing at a substantially smaller rate (RIN) or status quo (SQ).

We will use the alternative method shown by Fox (2015a) in order to create cardinal utilities for the payoff matrix. Another option would be the lottery method by von Neumann and Morgenstern (2004). First, we rank order the outcomes for each side in order of preference. These are the ordinal utilities.

First, we list the combinations of strategies that can be played in this game. This will be a 2 x 4 game. We also assume that this is a partial conflict (non-zero sum) game. We will refer to NS as R_1, S as R_2, SQ as C_1, In as C_2, RIN as C_3, and GIN as C_4.

$NS\ SQ \rightarrow R_1C_1$
$NS\ IN \rightarrow R_1C_2$
$NS\ GIN \rightarrow R_1C_3$
$NS\ RIN \rightarrow R_1C_4$
$S\ SQ \rightarrow R_2C_1$
$S\ IN \rightarrow R_2C_2$
$S\ RIN \rightarrow R_2C_3$
$S\ GIN \rightarrow R_2C_4$

8.5.2.1.2 Writer's Alternatives and Rankings

The writers have eight strategy combinations to rank. We start by ranking from best to worst. The writers prefer getting compensation to not getting any compensation.

$NS\ IN \rightarrow R_1C_2 = 8$
$NS\ GIN \rightarrow R_1C_3 = 7$
$NS\ RIN \rightarrow R_1C_4 = 6$
$S\ IN \rightarrow R_2C_2 = 5$
$S\ RIN \rightarrow R_2C_3 = 4$
$S\ GIN \rightarrow R_2C_4 = 3$
$NS\ SQ \rightarrow R_1C_1 = 2$
$S\ SQ \rightarrow R_2C_1 = 1$

8.5.2.1.3 Management's Alternatives and Rankings

The management prefers not to give additional compensation, but if they have to, they would prefer at a lower amount. Thus, management's ordinal ranking might be as follows:

$NS\ SQ \rightarrow R_1C_1 = 8$
$S\ SQ \rightarrow R_2C_1 = 7$
$NS\ RIN \rightarrow R_1C_4 = 6$

$S\ RIN \rightarrow R_2C_3 = 5$

$NS\ GIN \rightarrow R_1C_3 = 4$

$S\ GIN \rightarrow R_2C_4 = 3$

$NS\ IN \rightarrow R_1C_2 = 2$

$S\ IN \rightarrow R_2C_2 = 1$

Payoff matrix (ordinal)

		Management			
		C1	C2	C3	C4
Writers	R1	(2, 8)	(8, 2)	(7, 4)	(6, 6)
	R2	(1, 7)	(5, 1)	(4, 5)	(3, 3)

A movement diagram to find a pure strategy equilibrium. The pure strategy using the ordinal values is R1C1, which represents no strike and status quo. The management is quite happy, and the writers are very unhappy. As a matter of fact, the writers begin to consider to strike.

In order to continue the analysis with some accuracy, we need to convert the ordinal values into cardinal values. We could either use the lottery range using the method of von Neumann and Morgenstern (2004), Chapter 7, or the AHP method described by Fox (2015a). We illustrate the method described by using AHP within this analysis.

We began with analyzing the writer's strategies versus the management's strategies to obtain the ordinal ranking. Now we use the AHP method described in Chapter 4.

The AHP shows that the matrix, Figure 8.5, that is generated is consistent as the $CR = 0.00073$.

The output, the eigenvectors, are the cardinal values for our strategies. These are found for the writers:

R1C2	0.38079158
R1C3	0.22202891
R1C4	0.13584163
R2C2	0.09103997
R2C3	0.0604459
R2C4	0.04445166
R1C1	0.03563703
R2C1	0.02976333

We repeat the identical process for the management. The pairwise comparisons template is shown in the matrix in Figure 8.6.

Matrix O

	R1C2 (1)	R1C3 (2)	R1C4 (3)	R2C2 (4)	R2C3 (5)	R2C4 (6)	R1C1 (7)	R2C1 (8)
1 R1C2	1	2	3	4	6	7	8	9
2 R1C3	1/2	1	2	3	4	6	7	8
3 R1C4	1/3	1/2	1	2	3	4	5	6
4 R2C2	1/4	1/3	1/2	1	2	3	4	5
5 R2C3	1/6	1/4	1/3	1/2	1	2	3	4
6 R2C4	1/7	1/6	1/4	1/3	1/2	1	2	3
7 R1C1	1/8	1/7	1/5	1/4	1/3	1/2	1	2
8 R2C1	1/9	1/8	1/6	1/5	1/4	1/3	1/2	1

FIGURE 8.5
Pairwise comparison matrix for writers.

Matrix O

	R1C1 (1)	R2C1 (2)	R1C4 (3)	R2C3 (4)	R1C3 (5)	R2C4 (6)	R1C2 (7)	R2C2 (8)
1 R1C1	1	3	4	5	6	7	8	9
2 R2C1	1/3	1	3	4	4	6	7	8
3 R1C4	1/4	1/3	1	3	3	4	5	6
4 R2C3	1/5	1/4	1/3	1	3	3	4	5
5 R1C3	1/6	1/5	1/3	1/3	1	3	3	4
6 R2C4	1/7	1/6	1/4	1/3	1/3	1	3	3
7 R1C2	1/8	1/7	1/5	1/4	1/3	1/3	1	3
8 R2C2	1/9	1/8	1/6	1/5	1/4	1/3	1/3	1

FIGURE 8.6
Pairwise comparison matrix for management.

The output eigenvector, the cardinal values, for our strategies for the management are:

R1C1	0.42608847
R2C1	0.19934334
R1C4	0.11677389
R2C3	0.0807647
R1C3	0.0619435
R2C4	0.04651494
R1C2	0.03758504
R2C2	0.03098613

This provides us with a payoff matrix consisting of cardinal utilities, Table 8.2. This use of cardinal utilities is important because we really cannot do any mathematics (addition, subtraction, multiplication, or division) with ordinal values. Cardinal values will allow us to compute, as necessary. This will allow us to employ the cardinal values in the Nash arbitration scheme or employ them to see if any mixed strategy Nash equilibrium exists.

Nash (1950a) proved that every two-person game has at least one equilibrium either in Pure or in mixed (equalizing) strategies. The equilibriums are also called Nash equilibriums. He also developed the Nash arbitration (1950b) method which was use in this analysis.

8.5.2.2 Finding the Nash Equilibrium

For games with two players and more than two strategies each, we present the nonlinear optimization approach by Barron (2013). Consider a two-person game with a payoff matrix as before. Let's separate the payoff matrix into two matrices **M** and **N** for Players 1 and 2. We solve the following nonlinear optimization formulation in expanded form, in Equation (8.6).

$$\text{Maximiz} \sum_{i=1}^{n}\sum_{j=1}^{m} x_i a_{ij} y_j + \sum_{i=1}^{n}\sum_{j=1}^{m} x_i b_{ij} y_j + - p - q$$

TABLE 8.2

Payoff Matrix for the Game with Cardinal Values

		Management			
		C1	C2	C3	C4
Writers	R1	(0.0356, 0.426)	(0.38079, 0.0375)	(0.222, 0.0619)	(0.1358, 0.1167)
	R2	(0.02976, 0.0309)	(0.091, 0.03098)	(0.0604, 0.0807)	(0.0444, 0.04651)

Subject to:

$$\sum_{j=1}^{m} a_{ij} y_j \le p, \ i = 1, 2, \ldots, n,$$

$$\sum_{i=1}^{n} x_i b_{ij} \le q, \ j = 1, 2, \ldots, m, \tag{8.6}$$

$$\sum_{i=1}^{n} x_i = \sum_{j=1}^{m} y_j = 1$$

$$x_i \ge 0, y_j \ge 0$$

We used the computer algebra system Maple to input the game and then solve it. We let the matrix a_{ij} be labeled M and b_{ij} be labeled N in Maple. We wrote a short macro to perform the work.

The commands in Maple are:

- With (LinearAlgerba):with(Optimization)
- A:=Matrix([[0.0356,0.02976],[0.38079,0.091],[[0.222,0.0619],[0.135,0.0444]]);

$$A := \begin{bmatrix} 0.0356 & 0.02976 \\ 0.38079 & 0.091 \\ 0.222 & 0.0619 \\ 0.1358 & 0.0444 \end{bmatrix}$$

- B:=Matrix([[0.426,0.0309],[0.0375,0.03098],[0.0619,0.0807],[[0.1167,0.04651]]);

$$B := \begin{bmatrix} 0.426 & 0.0309 \\ 0.0375 & 0.03098 \\ 0.0619 & 0.0807 \\ 0.1167 & 0.04651 \end{bmatrix}$$

- $M:=Transpose(A):N:=Transpose(B):$
- $X:=`<,>`(x[1],x[2]):Y:=`<,>`(y[1],y[2],y[3],y[4]):$
- $c1:=seq(Transpose(X).N([j] < q, j=1..4):$
- $c2:=seq(M.Y)[j] < p, j=1..2):$
- $c3:=add(x[j],j=1..2)=1:$
- $c4:=add(y[j], j=1..4)=1:$
- $const:=\{c1,c2,c3,c4\};$

$const: = \{x_1 + x_2 = 1, y_1 + y_2 + y_3 + y_4 = 1, 0.0375x_1 +$
$0.03098x_2 \leq q, 0.0619x_1 + 0.0807x_2 \leq q, 0.1167x_1 +$
$0.04651x_2 \leq q, 0.426x_1 + 0.0309x_2 \leq q, 0.02971y_1 +$
$0.091y_2 + 0.0619y_3 + 0.0444y_4 \leq p, 0.0356y_1 +$
$0.38079y_2 + 0.222y_3 + 0.1358y_4 \leq p\}$

- *objective:=expand(Transpose(X).M.Y + Transpose(X).N.Y-p-q);*

$objective: = 0.4616y_1x_1 + 0.06066y_1x_2 + 0.41829y_2x_1 +$
$0.12198y_2x_2 + 0.2839y_3x_1 + 0.1426y_3x_2 +$
$0.2525y_4x_1 + 0.09091y_4x_2 - p - q$

- *QPSolve(objective, const, assume=nonnegative, maximize, iterationlimit=1000);*

$[2.06519568113350 \ 10^{-9}, [p = 0.0355999989674022, \ q = 0.425999998976,$
$y_1 = 1, \ y_2 = 0, y_3 = 0, y_4 = 0, x_1 = 1, x_2 = 0]]$

- QPSolve(objective, const, assume=nonnegative, maximize, intial point={p=0.07,q=.59});

$[0., [p = 0.3560000, \ q = 0.42600, y_1 = 1, \ y_2 = 0, y_3 = 0, y_4 = 0, x_1 = 1, x_2 = 0]]$

The Nash equilibrium, as expected is still (0.0356, 0.426) at R1C1. Changing the initial points did not uncover any additional equalizing strategy equilibriums. We also note that this result is not satisfying to the Writer's Guild and that they would like to have a better outcome.

We define the following terms:

Pareto Principle: "To be acceptable as a solution of the game, an outcome should be Pareto Optimal" from Straffin (2004).

Pareto Optimal: The outcome where neither player can improve payoff without hurting (decreasing the payoff) the other player.

As in this case, group rationality (Pareto) is sometimes in conflict with the individual rationality (dominant). The eventual outcome depends on the players. Obtaining a Pareto optimal outcome usually requires some sort of communication and cooperation among the players.

With the assumption that the outcome should be Pareto optimal, the next question is, "What is Pareto optimal, and what is it not (Pareto inferior)?" The simplest way for this to be understood is to draw a payoff polygon of the game. On the chart, the X-axis depicts the payoffs of Rose, and the Y-axis depicts the payoffs of Colin. By plotting the pure strategy solutions on the chart, one can see that the convex (everything inside) polygon enclosing

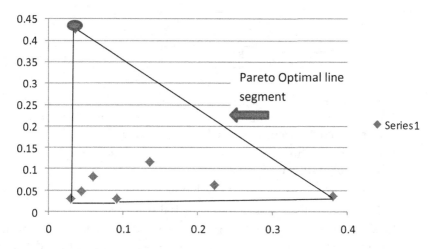

FIGURE 8.7
The payoff polygon.

the pure strategy solutions is then the payoff polygon or the feasible region. Therefore, the points inside the polygon are the possible solutions of the game.

Using Excel, we plot these coordinates from the payoff matrix to determine if any points are Pareto optimal. See Figure 8.7 for payoff polygon.

The Nash equilibrium at (0.0356, 0.426) lies along the Pareto optimal line segment. The writers are still unhappy with the result of this equilibrium and want arbitration, discussed in Chapter 11.

Chapter 8 Exercises

8.1 Corporation XYZ consists of companies Rose and Colin. Company Rose can make products R1 and R2. Company Colin can make products C1 and C2. These products are not in strict competition with one another, but there is an interactive effect depending on which products are on the market at the same time as reflected in the table below. The table reflects profits in millions of dollars per year. For example, if products R2 and C1 are produced and marketed simultaneously, Rose's profits are 3 million and Colin's are 4 million annually. Rose can make any mix of R1 and R2, and Colin can make any mix of C1 and C2. Assume the information below is known to each company.

		COLIN	
		C1	C2
ROSE	R1	(1, 5)	(8, 3)
	R2	(3, 4)	(2, 7)

a. Suppose the companies implement market strategies independently without communicating with one another. What do you think a likely outcome would be? Justify your choice.

b. In the event things turn "hostile" between Rose and Colin, find and explain the meaning of

1. Rose's security level and prudential strategy?

2. Colin's security level and prudential strategy?

Now suppose that the CEO is disappointed with the lack of spontaneous cooperation between Rose and Colin and decides to intervene and dictate the "best" solution for the corporation and employs an arbiter to determine an "optimal production and marketing schedule" for the corporation.

c. Explain the concept of "Pareto optimal" from the CEO's point of view. Is the "likely outcome" you found above Pareto optimal?

8.2 Solve the following partial conflict game for all its equilibria. Is any Pareto optimal?

		Colin	
		C1	C2
Rose	R1	(2, 0)	(1, 3)
	R2	(0, 1)	(3, 0)

8.3 Solve the following partial conflict game for all its equilibria. Is any Pareto optimal?

		Colin	
		C1	C2
Rose	R1	(2, 1)	(−1, −1)
	R2	(−1, −1)	(1, 2)

8.4 Solve the following partial conflict game for all its equilibria. Is any Pareto optimal?

		Colin		
		C1	C2	C3
Rose	R1	(−2, −4)	(3, −2)	(1, 4)
	R2	(−3, −3)	(2, 1)	(3, 4)
	R3	(2, 3)	(1, 1)	(3, −1)

References

Barron, E.N. (2013). *Game Theory: An Introduction*. Hoboken, NJ: John Wiley & Sons.

Feix, M. (2007). Game theory toolkit and workbook for defense analysis students, Master's Thesis, Naval Postgraduate School, Monterey, CA.

Fox, W.P. (2015a). An alternative approach to the lottery method in utility theory for game theory. *American Journal of Operations Research*, 5 (3). https://doi.org/10.4236/ajor.2015.53016.

Fox, W.P. (2015b). Analysis of the 2007-2008 Writer's Guild Strike with Game Theory. *Applied Mathematics Journal*, 6, pp. 2132–2141. http://file.scirp.org/pdf/AM_2015113010070339.pdf.

Giordano, F., W. Fox, & S. Horton. (2013). *A First Course in Mathematical Modeling*, 5th ed. Boston, MA: Brooks-Cole.

Nash, J.F. (1950a). Equilibrium points in n-person games. *Proceedings of the National Academy of Sciences of the United States of America*, 36, (1), pp. 48–49.

Nash, J.F. (1950b). The bargaining problem. *Econometrica*, 18, (2), pp. 155–162.

Straffin, P. (2004). *Game Theory and Strategy*. Washington, DC: Mathematical Association of America.

von Neumann, J. & O. Morgenstern. (2004). *Theory of Games and Economic Behavior*, 60th anniversary ed. Princeton, NJ: Princeton University Press.

Additional Readings

Chiappori, A., S. Levitt, & T. Groseclose. (2002). Testing mixed-strategy equilibria when players are heterogeneous: The case of penalty kicks in soccer. *American Economic Review*, 92 (4), pp. 1138–1151.

Couch, C., W.P. Fox, & S. Everton. (2015). Mathematical modeling and analysis of a dark money network. *Journal of Defense Modeling and Simulation*, pp. 1–12. https://doi.org/10.1177/1548512915625337.

Fox, W. (2008). Mathematical modeling of conflict and decision making: The writers' guild strike 2007–2008. *Computers in Education Journal*, 18 (3), pp. 2–11.

Satty, T. (1980). *The Analytical Hierarchy Process*. United States: McGraw Hill.

The New York Times. Who won the writers' strike? *The New York Times*. Retrieved on February 12, 2008, https://www.nytimes.com/2008/02/12/arts/television/12strike.html.

Wikipedia. (2007). The Writer's Guild Strike. http://en.wikipedia.org/wiki/2007_Writers_Guild_of_America_strike.

Writers Guild of America. (2007). WGA Contract 2007 Proposals (PDF). Retrieved November 9, 2007.

9

Evolutionary Stable Strategies

9.1 Introduction

The application of evolutionary stable strategies (ESS) was first introduced in 1973 (Maynard Smith and Price) and also by Straffin (2004). This was marked as a powerful explanatory idea in evolutionary biology. The behavioral and psychological patterns lead to an ESS.

In applied game theory, we assume all players are rational actors. That is, each player understands and knows the payoff matrix and each player knows how to the play game motivated by the maximum payout.

In biological aspects, not all actors comprehend the payout matrix, we consider biological resource competition and natural selection.

Thus, we will consider **Evolutionary Stable Strategies**.

ESS equilibrium is the strategy for behavior conflict between actors that are subject to evolutionary pressure rather than rational thought.

Let's demonstrate by an example.

Example 9.1: Hawks-Doves

Assumes that there is repeated random conflict over some resource. Conflict occurs between two players and only one player can win the resource. They do not share the resource.

Each player has only two possible strategies. Hawk must fight for resource. Dove must symbolically posture, but not fight. Winning the resource is worth 50 "fitness points".

Applying the points, we get the following:

Hawk versus Hawk
>Winner gets resource (50 points).
>Loser gets injured (–100 points).

Hawk versus Dove
>Hawk gets resource (50 points).
>Dove is uninjured (0 points).

Dove versus Dove
>Both spend a long time posturing.
>One eventually wins resource (50 points).
>Points deducted for wasted time (–10 points).

DOI: 10.1201/9781032726885-9

Our payoff matrix in general according to Maynard Smith and Price (1973) is as follows:

		Player 2	
		Hawk	Dove
Player 1	Hawk	$(v-c)/2, (v-c)/2$	$v,0$
	Dove	$0, v$	$v/2, v/2$

Letting $v = 50$ and $c = 100$

$$(v-c)/2 = -50/2 = -25$$

$v/2 = 25$ but each loses 10 points for wasted time and yields 15. The final payoff matrix, with values, is:

		Player 2	
		Hawk	Dove
Player 1	Hawk	$-25, -25$	$50, 0$
	Dove	$0, 50$	$25 - 10, 25 - 10$
			$15, 15$

Let's first consider a population that consists of ¼ Hawks and ¾ Doves.

		Player 2	
		Hawk	Dove
Player 1	Hawk	-25	50
	Dove	0	15

So, E[Player 1] = ¼ (−25) + 3/4 (50) = 125/4 = 31 ¼ versus ¼ (0) + ¾ (15) = 45/4 = 11 1/4

The maximum result is more Hawks.

Similarly, we can show if there are more Doves, then Doves increase. Therefore, it makes sense that there is a mixed number of Hawks and Doves in which neither has an advantage. We will call this result the ESS. In that case, our expected values for Hawks and Doves must be equal. We set the percentage of Hawks at x and for Doves at $1 - x$.

$$E[\text{Hawks}] = -25\,(x) + 50(1-x) = 50 - 75x$$

$$E[\text{Doves}] = x(0) + 15(1-x) = 15 - 15x$$

We set these equal and solve for x and $1 - x$.

$$50 - 75x = 15 - 15x$$

$$35 = 60x$$

GAME 1	A	B
A	1	2
B	3	4

GAME 2	A	B
A	3	1
B	2	4

GAME 3	A	B
A	1	4
B	2	3

- Game 1 demonstrates dominance. ESS is B.
- In Game 2, A and B can be ESS, depending on which is decided first.
- Game 3 is a split strategy, solved like Hawk-Dove discussed earlier.

	A	B
A	a	b
B	c	d

FIGURE 9.1
Other variations in ESS games.

$$X = 35/60 = 7/12$$

$$\text{and } 1 - x = 5/12$$

A mixed strategy of 7/12 Hawks and 5/12 Doves would be ESS.
If the entire population evolves to this strategy, then no other strategy could invade this population and prosper.

Other game variations are shown in Figure 9.1.

9.1.1 ESS Conditions

A is an ESS if $a > c$; or $a = c$ and $b => d$
  B is an ESS if $d > b$; or $d = b$ and $c => a$
Let's add a bully to our environment.

Bully enters the environment
  Bully conditions are

In any contest, show the initial fight.
Continue to fight if your opponent does not fight back.
If your opponent fights back, run away.

		Player Two		
		Hawk	Dove	Bully
	Hawk	-25	50	50
Player One	Dove	0	15	0
	Bully	0	50	25

Bully strategy dominates Dove strategy, so that Dove eventually dies out.

Game reduces to Hawks and Bullies, mixed strategy of ½ is the only ESS.

Now, let's add a retaliator to our environment.

Retaliator enters
Retaliator conditions

In any contest, behave like dove at first.

If persistently attacked, fight back with total strength.

		Player Two			
		Hawk	Dove	Bully	Retal
	Hawk	-25	50	50	-25
Player One	Dove	0	15	0	15
	Bully	0	50	25	0
	Retal	-25	15	50	15

Bourgeois enters the mix.

Bourgeois conditions

Be a Hawk on own territory.

Be a Dove on someone else's territory.

Assumes Bourgeois will fight half of each territory.

		Player Two				
		Hawk	Dove	Bully	Retal	Bourg
	Hawk	-25	50	50	-25	12.5
Player One	Dove	0	15	0	15	7.5
	Bully	0	50	25	0	25
	Retal	-25	15	50	15	-5
	Bourg	-12.5	32.5	25	-5	25

We have two different ESSs:

Retaliator with some Doves coexisting.

Bourgeois with some Bullies coexisting.

9.1.2 Summary

Evolutionary Stable Strategies Application

- Understanding evolution of aggression
- Understanding biological functions
- Understanding altruism and cooperation

Further Research

- Prisoner's dilemma and cooperation
- Simple sexual mating game

Retaliator Conditions

- In any contest, behave like dove at first.
- If persistently attacked, fight back with total strength.

Chapter 9 Exercises

9.1 Show that in the game that introduces Bullies into the environment that ½ Bullies and ½ Hawks are the ESS.

9.2 Find the ESS when retaliators are added to the environment.

9.3 Consider a game where a female species tries to get a male species to stick around and help raise a family. Let's assume the payoff matrix is as follows:

		Male	
		Faithful	Philandering
Female	Coy	(2, 2)	(0,0)
	Fast	(5, 5)	(−5, 15)

a. Find the pure strategy, if one exists. If not, explain why one does not exist. Use a movement diagram to assist you.

b. The ESS for males would be one that equalizes the expected value to coy and fast females. Find it.

c. Find the ESS for females.

d. If both males and females follow their respective ESS, what is the outcome?

References

Maynard Smith, J. & G. Price. (1973). The logic of animal conflict. *Nature,* (246), pp. 15–18.

Straffin, P. (2004). *Game Theory and Strategy.* Washington, DC: MAA Publications, pp. 93–100.

10

Communications in Partial Conflict Games

10.1 Introduction

We have only present games where each player chooses their strategies simultaneously and without communications. We refer to communications as any method to communicate your intentions. Many games in real life are not like this. In this chapter, we will consider things that might happen when players communicate. We will consider players who might take the initiative and move first, players who might threaten the other player, players who make promises to other players, and even a combination of threats and promises. We will consider these options as "strategic moves".

We will begin with the game of chicken and start with a review.

10.2 The Game of Chicken without Communication

Chicken, as we have discussed in our section on classic games, is the usual game used to model conflicts in which the players are on a collision course. The players may be drivers approaching each other on a narrow road, in which each has the choice of swerving to avoid a collision or not swerving. In the movie *Rebel Without a Cause*, starring James Dean, the drivers were two teenagers, but instead of bearing down on each other they both raced toward a cliff, with the object being not to be the first driver to slam on his brakes and thereby "chicken out", while, at the same time, not plunging over the cliff. Both players were expecting to bail out, the last to bail out being declared the winner. Another example was in the movie *Footloose*, staring Kevin Bacon, where he and his opponent are on tractors heading toward each other on a collision course. The last that veers away is the winner.

DOI: 10.1201/9781032726885-10

We again assume without communication each player will play his maximin strategy

	Colin		Rose: Row Minimum
	Swerve	**Not Swerve**	
Swerve	(3, 3) $\Longrightarrow$	(2, 4)	2
Rose	$\downarrow$	$\uparrow$	
Not Swerve	(4, 2) $\Longleftarrow$	(1, 1)	1
Colin: Column Minimum	2	1	

We see that Rose's maximin strategy is Swerve since she is guaranteed at least a 2 if she plays Swerve. Similarly, Colin's maximin strategy is Swerve. If both players play conservatively without communication, we would expect the result to be (Swerve, Swerve) with payoff (3, 3). Note from the movement diagram that neither player has a dominant strategy. Also note from the movement diagram that (3, 3) is not a Nash equilibrium, and either player can improve unilaterally by switching to her or his Not Swerve strategy. However, if both players switch, then the result is the disaster outcome (Not Swerve, Not Swerve) with payoffs (1, 1). There are two Nash equilibriums with payoffs (4, 2) and (2, 4). Now imagine a confrontation between two nations, such as the Cuba Missile Crisis which is discussed in Section 10.5. Both countries are sitting on Swerve, but have the motivation to switch to Not Swerve at the last minute. If both switch, a catastrophe occurs.

Summarizing the game of Chicken,

Chicken is a two-person partial conflict game in which each player has two strategies: Swerve to avoid a crash or Not Swerve to attempt to win the game. Neither player has a dominant strategy. If both players choose Swerve, the outcome is not a Nash equilibrium and therefore unstable. There are two Nash equilibriums where one of the two players chooses Swerve and the second player chooses Not Swerve.

The game of chicken is used to model such topics as the confrontations between countries or species.

10.3 The Game of Chicken with Communication

10.3.1 Moving First or Committing to Move First

We now assume both players can communicate their plans or their moves to the second player. If Rose can move first, she can choose Swerve or

Not Swerve. Examining the movement diagram, she should expect Colin's responses as follows:

If Rose plays Swerve, Colin plays Not Swerve resulting in the outcome (2, 4).

If Rose plays Not Swerve, Colin plays Swerve resulting in the outcome (4, 2).

Since Rose's objective is to maximize her outcome, she should choose to move first and communicate that she has selected Not Swerve. Colin's best response then is to Swerve resulting in (4, 2). She wins the duel. Of course, we must have a situation where Rose can move first. Or, she can communicate a commitment to the strategy Not Swerve, which, if made credible, again yields (4, 2), her best outcome.

10.3.2 Issuing a Threat

If Colin has the opportunity to move first or is committed to (or possibly considering) Not Swerve, Rose may have a threat to deter Colin from playing Not Swerve. Such a threat must satisfy three conditions:
Conditions for a threat by Rose

1. Rose communicates that she will play a certain strategy contingent upon a previous action of Colin.
2. Rose's action is harmful to Rose.
3. Rose's action is harmful to Colin.

In the game of Chicken, Rose wants Colin to play Swerve. Therefore, she makes the threat to Colin Not Swerve to deter him from choosing that strategy. Examining the movement diagram,

Normally, if Colin plays Not Swerve, Rose plays Swerve yielding (2, 4). In order to harm herself, Rose must play Not Swerve. Thus, the potential threat must take the form. If Colin plays Not Swerve, then Rose plays Not Swerve yielding (1, 1).

Is it a threat? It is contingent upon Colin choosing Not Swerve. Comparing (2, 4) and (1, 1), we see that the threat is harmful to Rose and is harmful to Colin. It is a threat and effectively eliminates the outcome (2, 4) making the game.

		Colin	
	Swerve		Not Swerve
Swerve	(3, 3)		Eliminated by threat
Rose	↓		
Not Swerve	(4, 2)	⟸	(1, 1)

Colin still has a choice of choosing Swerve or Not Swerve. Using the movement diagram, he analyzes his choices as follows:

If Colin selects Swerve, Rose chooses Not Swerve yielding (4, 2).

If Colin chooses Not Swerve, Rose chooses Not Swerve yielding (1, 1) (because of Rose's threat).

Thus, Colin's choice is between a payoff of 2 and 1. He should choose Swerve yielding (4, 2). If Rose can make her threat credible, she can secure her best outcome.

10.3.3 Issuing a Promise

Again, if Colin has the opportunity to move first or is committed to (or possibly considering) Not Swerve, Rose may have a promise to encourage Colin to play Swerve instead. A promise must satisfy three conditions:

Conditions for a Promise by Rose

1. Rose communicates that she will play a certain strategy contingent upon a previous action of Colin.
2. Rose's action is harmful to Rose.
3. Rose's action is beneficial to Colin.

In the game of Chicken, Rose wants Colin to play Swerve. Therefore, she makes the promise to Colin Swerve to sweeten the pot so he will choose Swerve. Examining the movement diagram,

Normally, if Colin plays Swerve, Rose plays Not Swerve yielding (4, 2). In order to harm herself, she must play Swerve. Thus, the promise takes the form. If Colin plays Swerve, then Rose plays Swerve yielding (3, 3).

Is it a promise? It is contingent upon Colin choosing Swerve. Comparing the normal (4, 2) with the promised (3, 3), we see that the promise is harmful to Rose and is beneficial to Colin. It is a promise and effectively eliminates the outcome (4, 2) making the game.

	Colin	
	Swerve	**Not Swerve**
Swerve	(3, 3) $\Longrightarrow$	(2, 4)
Rose		$\uparrow$
Not Swerve	Eliminated by Promise	(1, 1)

Colin still has a choice of choosing Swerve or Not Swerve. Using the movement diagram, he analyzes his choices as follows:

If Colin selects Swerve, Rose chooses Swerve yielding (3, 3) as promised.

If Colin chooses Not Swerve, Rose chooses Swerve yielding (2, 4).

Thus, Colin's choice is between payoffs of 3 and 4. He should choose Not Swerve yielding (2, 4). Rose does have a promise. But her goal is for Colin to choose Swerve. Even with the promise of eliminating an outcome, Colin chooses Not Swerve. The promise does not work. In the exercises, you are asked to show that if Rose and Colin both make a promise, then (3, 3) is the outcome.

In summary, the game of Chicken offers many options. If the players choose conservatively without communication, the maximin strategies yield (3, 3), which is unstable: both players unilaterally can improve their outcomes. If either player moves first or commits to move first, they can obtain their best outcome. For example, Rose can obtain (4, 2) which is a Nash equilibrium. If Rose issues a threat, she can eliminate (2, 4) and obtain (4, 2). A promise by Rose eliminates (4, 2) but results in (2, 4) which does not improve the (3, 3) likely outcome without communication.

Example 10.1: A Combination of Threat and Promise

Consider the following game:

	Colin		
	C1		**C2**
R1	(2, 4)	⟵	(3, 3)
Rose	↑		↓
R2	(1, 2)	⟵	(4, 1)

In the exercises, you are asked to show that without communication, if both players play their maximin strategies, the outcome is (2, 4), a Nash equilibrium, and that Colin has a dominant strategy C1. Without communication, Colin gets his best outcome, but can Rose do better that (2, 4) with a strategic move?

Rose first, if Rose moves R1, Colin should respond with C1 yielding (2, 4). If Rose moves R2, Colin responds with C1 yielding (1, 2). Rose's best choice is (2, 4), no better than the likely conservative outcome without communication.

Rose threat, Rose wants Colin to play C2. Normally, if Colin plays C1, Rose plays R1 yielding (2, 4). To hurt herself she must play R2 yielding (1, 2). Comparing the normal (2, 4) and (1, 2), the threat is contingent upon

Colin playing C1, which hurts Rose and hurts Colin. It is a threat and effectively eliminates (2, 4) yielding.

	Colin	
	C1	**C2**
R1	eliminated	(3, 3)
Rose		↓
R2	(1, 2) ⟸	(4, 1)

Does the threat deter Colin from playing C1? Examining the movement diagram, if Colin plays C1, the outcome is (1, 2). If Colin plays C2, the outcome is (4, 1). Colin's best choice is still C1. Thus, there is a threat, but it does not work. Does Rose have a promise that works by itself?

Rose promise, Rose wants Colin to play C2. Normally, if Colin plays C2, Rose plays R2 yielding (4, 1). To hurt herself, she must play R1 yielding (3, 3). Comparing the normal (4, 1) with the promised (3, 3), the move is contingent upon Colin playing C2, which hurts Rose and is beneficial to Colin. It is a promise and effectively eliminates (4, 1) yielding.

	Colin	
	C1	**C2**
R1	(2, 4) ⟸	(3, 3)
Rose	↑	
R2	(1, 2)	eliminated

Does the promise motivate Colin to play C2? Examining the movement diagram, if Colin plays C1, the outcome is (2, 4). If Colin plays C2, the outcome is (3, 3). Colin's best choice is still C1 for (2, 4). Thus, there is a promise, but it does not work. What about combining both the threat and the promise?

Combination of threat and promise, we see that Rose does have a threat that eliminates an outcome but does not work by itself. She also has a promise that eliminates an outcome but does not work by itself. In such situations, we can examine issuing both the threat and the promise to eliminate two outcomes to determine if a better outcome results. Rose's threat eliminates (2, 4), and Rose's promise eliminates (4, 1). If she issues both the threat and the promise, the following outcomes are available.

	Colin	
	C1	**C2**
R1	eliminated	(3, 3)
Rose		
R2	(1, 2)	eliminated

If Colin plays C1, the result is (1, 2), and choosing C2 yields (3, 3). He should choose C2, and (3, 3) represents an improvement for Rose over the likely outcome without communication (2, 4).

10.4 Credibility

Of course, commitments to first moves, threats, and promises must be made credible. If Rose issues a threat, and Colin chooses to Not Swerve anyway, will Rose carry out her threat and crash (1, 1) even though that action no longer promises to get her the outcome (4, 2)? If Colin believes that she will not carry through on her threat, he will ignore the threat. In the game of Chicken, if Rose and Colin both promise to Swerve and Colin believes Rose's promise and executes Swerve, will Rose carry out her promise to Swerve and accept (3, 3) even though (4, 2) is still available to her? One method for Rose to gain credibility is to lower one or more of her payoffs so that it is obvious to Colin that she will execute the stated move. Or, if possible, she may make a *side payment* to Colin to increase his selected payoffs in order to entice him to a strategy that is favorable to her and is now favorable to him because of the side payment. These ideas are pursued further in the exercises.

An inventory of the strategic moves available to each player is an important part of determining how a player should act. Each player wants to know what strategic moves are available to each of them. For example, if Rose has a first move and Colin has a threat, Rose will want to execute her first move before Colin issues his threat. The analysis requires knowing the rank order of the possible outcomes for both players. Once a player has decided which strategy he wants the opposing player to execute, he can then determine how the player will react to any of his moves. You may find Table 10.1 useful in learning how to organize the questions and responses that are necessary to determine the strategic moves available to each player.

10.4.1 The Battle of the Sexes

The third classical two-person partial conflict game is known as the Battle of the Sexes. The scenario is that He prefers to go to boxing and She prefers to go to ballet. They do prefer to go to the same event rather than go to an event alone. His ranking from best to worst is that (4) they both go to boxing, (3) they both go to ballet, (2) he goes to boxing and she goes to ballet, and (1) he goes to ballet and she goes to boxing.

	Boxing	She	Ballet
	C1		C2
Boxing	(4, 3)	⟸	(2, 2)
He	↑		↓
Ballet	(1, 1)	⟹	(3, 4)

In the exercises, you are asked to show that if this game is played conservatively without communication, he is likely to go to Boxing and she to Ballet

TABLE 10.1

Analysis for Strategic Moves

Simultaneous Without Communication
If both players maximize their outcomes the likely outcome is (__, __)
With Communication (Strategic Moves) from Rose's Perspective
FIRST MOVES
Should Rose move first:
If Rose does R1, then Colin does __, implies outcome (__, __)
      If Rose does R2, then Colin does __, implies outcome (__, __)
So Rose would choose outcome (__, __)
   Should Rose **force** Colin to move first:
If Colin does C1, then Rose does __, implies (__, __)
      If Colin does C2, then Rose does __, implies (__, __)
      So Colin would choose (__, __)
    Conclusions: Rose moving first would result in outcome (__, __)
                Forcing Colin to move first would result in outcome (__, __)
THREATS: Example: Suppose Rose wants Colin to play C2
If Colin does C1 and Rose does the opposite of what she logically should do (in order to hurt
  herself), then Rose does, __ with outcome (__, __)
      Does it also hurt Colin? If so, it is a threat and eliminates outcome (__, __)
With the threat implemented, Colin chooses __ and the outcome is (__, __)
Does the threat work alone? (Does she in fact get Colin C2?)
PROMISES: Example: Suppose Rose wants Colin to play C2
If Colin does C2 and Rose hurts herself, she does Rose __with outcome (__, __)
Does it help Colin? If so, it is a promise and eliminates (__, __)
With the promise implemented, the outcome is (__, __)
Does the promise work alone? (Does she in fact get Colin C2?)
COMBINATION THREAT AND PROMISE
Threat eliminates (__. __) **AND** the Promise eliminates (__. __)
Logical outcome is (__. __)
Summary of Strategic Moves available to Rose (and to Colin)

From Notes by Frank R. Girdano

with payoffs (2, 2). Each has a first move that gets them their best outcome. (He calls and says that his phone is about dead and he is headed to boxing. This move works once and is not good for repeated play!) Neither has a threat nor a promise. Obviously, arbitration which mixes the (Boxing, Boxing) and (Ballet, Ballet) outcomes is needed!

In this section, we have introduced the classical two-person games of partial conflict. Each of the games has many applications. For that reason, it is important to know how the games should be played. Would you prefer to play without communication? If communication is appropriate, what strategic moves do you and the opposing player have?

10.5 Classical Game Theory and the Missile Crisis

Game theory is a branch of mathematics concerned with decision-making in social interactions. It applies to situations (*games*) where there are two or more people (called *players*) each attempting to choose between two more ways of acting (called *strategies*). The possible outcomes of a game depend on the choices made by all players and can be ranked in order of preference by each player.

In some two-person, two-strategy games, there are combinations of strategies for the players that are in a certain sense "stable". This will be true when neither player, by departing from its strategy, can do better. Two such strategies are together known as a Nash equilibrium, named after John Nash, a mathematician who received the Nobel Prize in economics in 1994 for his work on game theory. Nash equilibrium does not necessarily lead to the best outcomes for one, or even both, players. Moreover, for the games that will be analyzed – in which players can only rank outcomes ("ordinal games") but not attach numerical values to them ("cardinal games") – they may not always exist. (While they always exist, as Nash showed, in cardinal games, Nash equilibrium in such games may involve "mixed strategies", which will be described later.). The following Cuban missile crisis model comes from Brams (1994).

Problem Identification Statement: Build a mathematical model that allows for consideration of alternative decisions by each side.

The Cuban missile crisis was precipitated by a Soviet attempt in October 1962 to install medium-range and intermediate-range nuclear-armed ballistic missiles in Cuba that were capable of hitting a large portion of the United States. The goal of the United States was the immediate removal of the Soviet missiles, and U.S. policymakers seriously considered two strategies to achieve this end (see Figure 10.1):

1. **A naval blockade (B)**, or "quarantine" as it was euphemistically called, to prevent the shipment of more missiles, possibly followed by stronger action to induce the Soviet Union to withdraw the missiles already installed.

2. **A "surgical" air strike (A)** to wipe out the missiles already installed, insofar as possible, perhaps followed by an invasion of the island.

		Soviet Union (S.U.)	
		Withdrawal (W)	Maintenance (M)
United States (U.S.)	Blockade (B)	Compromise (3,3)	Soviet victory, U.S. defeat (2,4)
	Air strike (A)	U.S. victory, Soviet defeat (4,2)	Nuclear war (1,1)

FIGURE 10.1
Cuban missile crisis as Chicken. *Key:* (x, y) = (payoff to U.S., payoff to S.U.), 4 = best, 3 = next best, 2 = next worst, and 1 = worst, Nash equilibrium underscored. Note there are two (4, 2) and (2, 4).

The alternatives open to Soviet policymakers were:

1. **Withdrawal (W)** of their missiles.
2. **Maintenance (M)** of their missiles.

These strategies can be thought of as alternative courses of action that the two sides, or "players" in the jargon of game theory, can choose. They lead to four possible outcomes, which the players are assumed to rank as follows: 4 = best, 3 = next best, 2 = next worst, and 1 = worst. Thus, the higher the number, the greater the payoff; but the payoffs are only *ordinal,* that is, they indicate an ordering of outcomes from best to worst, not the degree to which a player prefers one outcome over another. The first number in the ordered pairs for each outcome is the payoff to the row player (United States), the second number the payoff to the column player (Soviet Union).

As in chicken, as both players attempt to get to their equilibrium, the outcome of the games ends up at (1, 1). This is disastrous for both countries and their leaders. The best solution is the (3, 3) compromise position. However, (3, 3) is not stable. This will eventually put us back at (1, 1).

In this situation, one way to avoid this chicken dilemma is to try strategic moves.

Both sides did not choose their strategies simultaneously or independently. Soviets responded to our blockade after it was imposed. The United States held out the chance of an air strike as a viable choice even after the blockade. If the USSR would agree to remove the weapons from Cuba, the United States would agree to (a) remove the quarantine and (b) agree not to invade Cuba. If the Soviets maintained their missiles, the United States preferred the air-strike to the blockade. Attorney General Robert Kennedy said, "if they do not remove the missiles, then we will".

The United States used a combination of promises and threats. The Soviets knew their credibility in both areas was high (strong resolve). Therefore, they withdrew the missiles, and the crisis ended. Khrushchev and Kennedy were wise.

Theodore Sorensen, special counsel to Kennedy, used the language of moves to describe the actual deliberations within the executive Committee of key advisors to Kennedy. "We discussed what the Soviets reaction would be to any of our possible moves, what our reaction with them would have to be that Soviet action, and so on, trying to follow these roads to their ultimate conclusion".

Needless to say, the strategy choices, probable outcomes, and associated payoffs shown in Figure 10.1 provide only a skeletal picture of the crisis as it developed over a period of 13 days. Both sides considered more than the two alternatives listed, as well as several variations on each. The Soviets, for example, demanded the withdrawal of American missiles from Turkey as a *quid pro quo* for the withdrawal of their own missiles from Cuba, a demand publicly ignored by the United States.

Nevertheless, most observers of this crisis believe that the two superpowers were on a collision course, which is actually the title of one book describing this nuclear confrontation. They also agree that neither side was eager to take any irreversible step, such as one of the drivers in Chicken might do by defiantly ripping off the steering wheel in full view of the other driver, thereby foreclosing the option of swerving.

Although in one sense the United States "won" by getting the Soviets to withdraw their missiles, Premier Nikita Khrushchev of the Soviet Union at the same time extracted from President Kennedy a promise not to invade Cuba, which seems to indicate that the eventual outcome was a compromise of sorts. But this is not game theory's prediction for Chicken, because the strategies associated with compromise do not constitute Nash equilibrium.

To see this, assume the play is at the compromise position (3, 3), that is, the U.S. blockades Cuba and the S.U. withdraws its missiles. This strategy is not stable because both players would have an incentive to defect to their more belligerent strategy. If the United States were to defect by changing its strategy to the airstrike, the play would move to (4, 2), improving the payoff the United States received; if the S.U. were to defect by changing its strategy to maintenance, the play would move to (2, 4), giving the S.U. a payoff of 4. (This classic game theory setup gives us no information about which outcome would be chosen because the table of payoffs is symmetric for the two players. This is a frequent problem in interpreting the results of a game theoretic analysis, where more than one equilibrium position can arise.) Finally, should the players be at the mutually worst outcome of (1, 1), that is, nuclear war, both would obviously desire to move away from it, making the strategies associated with it, like those with (3, 3), unstable.

10.5.1 Theory of Moves and the Missile Crisis

Using Chicken to model a situation such as the Cuban missile crisis is problematic not only because the (3, 3) compromise outcome is unstable but also because, in real life, the two sides did not choose their strategies simultaneously, or independently of each other, as assumed in the game of Chicken described above. The Soviets responded specifically to the blockade after it was imposed by the United States. Moreover, the fact that the United States held out the possibility of escalating the conflict to at least an air strike indicates that the initial blockade decision was not considered final – that is, the United States considered its strategy choices still open after imposing the blockade.

As a consequence, this game is better modeled as one of *sequential bargaining*, in which neither side made an all-or-nothing choice but rather both considered alternatives, especially should the other side fail to respond in a manner deemed appropriate. In the most serious breakdown in the nuclear deterrence relationship between the superpowers that had persisted from the Second World War until that point, each side was gingerly feeling its way, step by ominous step. Before the crisis, the Soviets, fearing an invasion of Cuba by the United States and also the need to bolster their international strategic position, concluded that installing the missiles was worth the risk. They thought that the United States, confronted by a *fait accompli*, would be deterred from invading Cuba and would not attempt any other severe reprisals. Even if the installation of the missiles precipitated a crisis, the Soviets did not reckon the probability of war to be high (President Kennedy estimated the chances of war to be between 1/3 and 1/2 during the crisis), thereby making it rational for them to risk provoking the United States.

There are good reasons to believe that U.S. policymakers did not view the confrontation to be Chicken-like, at least as far as they interpreted and ranked the possible outcomes. We offer an alternative representation of the Cuban missile crisis in the form of a game we will call *Alternative*, retaining the same strategies for both players as given in Chicken but presuming a different ranking and interpretation of outcomes by the United States (see Figure 10.2). These rankings and interpretations fit the historical record better than those of "Chicken", as far as can be told by examining the statements made at the time by President Kennedy and the U.S. Air Force, and the type and number of nuclear weapons maintained by the S.U. (more on this below).

1. **BW**: The choice of blockade by the United States and withdrawal by the Soviet Union remains the compromise for both players – (3, 3).
2. **BM**: In the face of a U.S. blockade, Soviet maintenance of their missiles leads to a Soviet victory (its best outcome) and U.S. capitulation (its worst outcome) – (1, 4).

		Soviet Union (S.U.)		
		Withdrawal (W)		Maintenance (M)
United States (US)	Blockade (B)	**Compromise (3,3)**	→	Soviet victory, U.S. capitulation (1,4)
		↑		↓
	Air strike (A)	"Dishonorable" U.S. action, Soviets thwarted (2,2)	←	"Honorable" U.S. action, Soviets thwarted (4,1)

FIGURE 10.2
Cuban missile crisis as Alternative. *Key:* (x, y) = (payoff to U.S., payoff to S.U.), 4 = best, 3 = next best, 2 = next worst, and 1 = worst, the equilibrium is in bold, arrows indicate the movement diagram.

3. **AM**: An air strike that destroys the missiles that the Soviets were maintaining is an "honorable" U.S. action (its best outcome) and thwarts the Soviets (their worst outcome) – (4, 1).

4. **AW**: An air strike that destroys the missiles that the Soviets were withdrawing is a "dishonorable" U.S. action (its next-worst outcome) and thwarts the Soviets (their next-worst outcome) – (2, 2).

Even though an air strike thwarts the Soviets at both outcomes (2, 2) and (4, 1), I interpret (2, 2) to be less damaging for the Soviet Union. This is because world opinion, it may be surmised, would severely condemn the air strike as a flagrant overreaction – and hence a "dishonorable" action by the United States – if there were clear evidence that the Soviets were in the process of withdrawing their missiles anyway. On the other hand, given no such evidence, a U.S. air strike, perhaps followed by an invasion, would action to dislodge the Soviet missiles.

The statements of U.S. policymakers support Alternative. In responding to a letter from Khrushchev, Kennedy said,

> If you would agree to remove these weapons systems from Cuba ... we, on our part, would agree ... (a) to remove promptly the quarantine measures now in effect and (b) to give assurances against an invasion of Cuba,

which is consistent with Alternative since (3, 3) is preferred to (2, 2) by the United States, whereas (4, 2) is not preferred to (3, 3) in Chicken.

If the Soviets maintained their missiles, the United States preferred an air strike to the blockade. As Robert Kennedy, a close adviser to his brother during the crisis, said,

> If they did not remove those bases, we would remove them,

which is consistent with Alternative, since the United States prefers (4, 1) to (1, 4) but not (1, 1) to (2, 4) in Chicken.

Finally, it is well known that several of President Kennedy's advisers felt very reluctant about initiating an attack against Cuba without exhausting less belligerent courses of action that might bring about the removal of the missiles with less risk and greater sensitivity to American ideals and values. Pointedly, Robert Kennedy claimed that an immediate attack would be looked upon as "a Pearl Harbor in reverse, and it would blacken the name of the United States in the pages of history", which is again consistent with the Alternative since the United States ranks AW next worst (2) – a "dishonorable" U.S. action – rather than best (4) – a U.S. victory – in Chicken.

If Alternative provides a more realistic representation of the participants' perceptions than Chicken does, standard game theory offers little help in explaining how the (3, 3) compromise was achieved and rendered stable. As in Chicken, the strategies associated with this outcome are not a Nash equilibrium, because the Soviets have an immediate incentive to move from (3, 3) to (1, 4).

However, unlike Chicken, Alternative has no outcome at all that is a Nash equilibrium, except in "mixed strategies". These are strategies in which players randomize their choices, choosing each of their two so-called pure strategies with specified probabilities. But mixed strategies cannot be used to analyze Alternative, because to carry out such an analysis, there would need to be numerical payoffs assigned to each of the outcomes, not the rankings I have assumed.

The instability of outcomes in Alternative can most easily be seen by examining the cycle of preferences, indicated by the arrows going in a clockwise direction in this game. Following these arrows shows that this game is *cyclic*, with one player always having an immediate incentive to depart from every state: the Soviets from (3, 3) to (1, 4), the United States from (1, 4) to (4, 1), the Soviets from (4, 1) to (2, 2), and the United States from (2, 2) to (3, 3). Again we have indeterminacy, but not because of multiple Nash equilibrium, as in Chicken, but rather because there is no equilibrium in pure strategies in Alternative.

Chapter 10 Exercises

Find all the pure strategy solutions in Problems 10.1–10.6. Then use strategic moves (using Table 10.1) to determine if Rose can get a better outcome.

10.1

		Colin	
		C_1	C_2
Rose	R_1	(2, 3)	(3, 1)
	R_2	(1, 4)	(4, 2)

10.2

		Colin	
		C_1	C_2
Rose	R_1	(1, 2)	(3, 1)
	R_2	(2, 4)	(4, 3)

10.3

		Colin	
		C_1	C_2
Rose	R_1	(2, 2)	(4, 1)
	R_2	(1, 4)	(3, 3)

10.4

		Colin	
		C_1	C_2
Rose	R_1	(2, 6)	(10, 5)
	R_2	(4, 8)	(0, 0)

10.5 Consider the following 2 × 2 partial conflict game, and find the solution when there is no communication between the players.

		Colin	
		C1	C2
Rose	R1	(2, 4)	(1, 0)
	R2	(3, 1)	(0, 0)

10.6 Given the payoff matrix below from the battle of the sexes:

		Colin	
		C1	C2
Rose	R1	(4, 3)	(2, 2)
	R2	(1, 1)	(3, 4)

If there is no communication between Rose and Colin, find the solution. Determine, if we have communications and employ strategic moves, what possible outcomes are available.

Projects

10.1 Consider the following 2 × 2 partial conflict game

		Colin	
		C1	C2
Rose	R1	(3, 5)	(0, 1)
	R2	(6, 2)	(−1, 4)

Find the solution assuming no communications.
Apply strategic moves to see if we can improve our outcomes.

10.2 Find the present classical Prisoner's Dilemma concerning suspects in a crime.

10.3 Create your own scenario that follows either the format of the game of Battle of the Sexes, Chicken, or the Prisoner's Dilemma. Identify all strategies, and use ordinal ranking to give them value. Completely solve your game.

Reference

Brams, S.J. (1994). *Theory of Moves*. Cambridge, UK: Cambridge University Press.

Additional Readings

Brams, S.J. (1997). Game theory and emotions. *Rationality and Society,* 9 (1), pp. 93–127.
Brams, S.J. (1999). Modeling free choice in games. in Edited by Myrna H. Wooders. *Topics in Game Theory and Mathematical Economics: Essays in Honor of Robert J. Aumann.* Providence, RI: American Mathematical Society, pp. 41–62.
Brams, S.J. & C.B. Jones. (1999). Catch-22 and King-of-the-Mountain Games: Cycling, frustration and power. *Rationality and Society,* 11 (2), pp. 139–167.
Giordano, F. (2006), Notes form DA 4410 Course. Monterey, CA.
Wilson, S.J. (1998). Long-term behaviour in the Theory of Moves. *Theory and Decision,* 45 (3), pp. 201–240.

11

Nash Arbitration Method

11.1 Introduction to Nash Arbitration

Let's start with Nash's arbitration theorem (1950a, 1950b),

> There is one and only one arbitration scheme that satisfies Axioms 1 through Axiom 4. It is this: if $SQ=(x_0, y_0)$, then the arbitrated solution point N is the point (x, y) in the polygon with $x \geq x_0, y \geq y_0$ which maximizes the product $(x-x_0)(y-y_0)$.

We define the status quo point as the security levels found by optimizing the matrix games for our two players. The status quo solution for each player in their game is their security level. We illustrate this later in our examples.

The Axioms 1–4 mentioned above are (Nash, 1950a, 1950b; Straffin, 2004):

Axiom 1. Rationality

Axiom 2. Linear Invariance

Axiom 3. Symmetry

Axiom 4. Independence of Irrelevant Alternatives

More information can be found on these axioms and the arbitration scheme in Nash (1950a, 1950b) and Straffin (2004).

Our goal is to have you understand what the method does, how to obtain solutions, and how to "play" the arbitration game using algebraic concepts in order to obtain the solution that is fairest to both players and being Pareto optimal.

What does it mean to be Pareto optimal? In game theory, major assumptions are that the games are simultaneous, repetitive, players have perfect knowledge, and the players act rationally. Rationality of players implies that each player wants the best solution possible versus their opponent. For a more thorough discussion of rationality, see Straffin (2004) for his arguments. If any of these assumptions are not valid then introductory game theory methods cannot be used.

DOI: 10.1201/9781032726885-11

For the players, we place Player 1 on the horizontal axis and Player 2 on the vertical axis. The northeast region is best for both players. The northeast corner is where both x and y are each the greatest within the constraints of the strategies as we will show. We plot the points of the game's strategies on the axis and connect the outer boundary of points to keep the polygon convex. The northeast region of this polygon is called the Pareto optimal line segment. We provide a graphical depiction of this later in this article.

11.2 Methods without Calculus

To solve the Nash arbitration scheme, we maximize $Z = (x - x_0) * (y - y_0)$.
We can employ several methods to find this point, without calculus.
We start by knowing the end points of the Pareto optimal line segment then (x_1, y_1) and (x_2, y_2) then these methods can be used.

Step 1. Find the equation of the line passing through the end points of the Pareto optimal line. Use the point-slope formula to obtain $y = mx + b$.

Step 2. Substitute $y = mx + b$ for y to obtain $(x - x_0)*(mx + b - y_0)$.

Issues: In our first example, we do not know the end points of the Pareto optimal line. So we need a method to assist us. For all points on the Pareto frontier (the NE region), we can substitute the points from left most point to right most into the equation $(x - x_0) * (y - y_0)$ and obtain the values. The largest two values provide the end points to the Pareto optimal line segment. We can use those end points in the following two methods as we will show.

Example 11.1: Consider the Following Matrix Game

		Colin	
		C1	C2
Rose	R1	(0, 5)	(3, 4)
	R2	(1, 0)	(6, 1)

The vertices along the Pareto frontier in Figure 11.1 are (0, 5), (3, 4), and (6, 1). We use (1, 1) as our status quo point.

Points	$(x - 1) * (y - 1)$
(0, 5)	−4
(3, 4)	6
(6, 1)	0

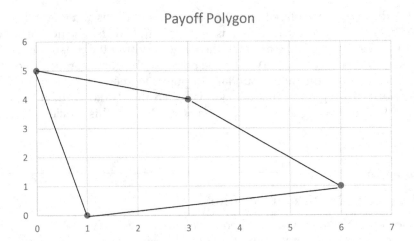

FIGURE 11.1
Payoff polygon.

Therefore, the end points (vertices) for the Pareto optimal line segment are (3, 4) and (6, 1).

Method 1. The Giordano Geometry Method, Naval Postgraduate School (NPS)

The use of midpoints is required.

> **Step 1.** Draw a horizontal and vertical line through the status quo points until it intersects the line for the Pareto optimal line.
> **Step 2.** Find the coordinates of these intersections.
> **Step 3.** Compute the x and y midpoints.
> **Step 4.** If either the x or y midpoint is outside the domain or range, then choose the closest end point of the Pareto optimal line segment. Otherwise, the midpoints are the Nash arbitration point. Call these Nash arbitration coordinates (x^*, y^*).
> **Step 5.** Compute Z the value of $(x^* - x_0)(y^* - y_0)$.
> The Pareto optimal line is $y - 4 = -(x - 3)$ or $y = -x + 7$.

The coordinates of the triangle are the status quo point (1, 1), (6, 1), and (1, 6) as shown in Figure 11.2.
We find the midpoints:

$$y_{midpoint} = 3.5$$
$$x_{midpoint} = 3.5$$

These midpoints are the Nash arbitration point, (3.5, 3.5), and the value of $Z = (x - x_0) * (y - y_0)$ is 6.25.

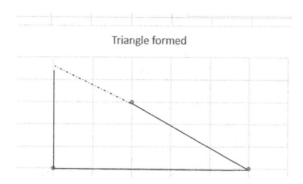

Triangle formed

FIGURE 11.2
Our method's triangle.

Next, we find that the probabilities to play the strategies R1C2 and
R2C2 are found by solving two equations and two unknowns as we
stated before.

$$3p_1 + 6p_2 = 3.5$$
$$4p_1 + 1p_2 = 3.5$$

We solve the two equations and two unknowns to find the probabilities.

$$p_1 = 5/6$$
$$p_2 = 1/6$$

Knowing Colin always plays C2 means that Rose plays strategy R1
five-sixths of the time and strategy R2 only one-sixth of the time during
the arbitration.

Method 2. Parabolic Method (also without Calculus)

As shown in Figure 11.3, we see a parabola that might be typical in these
types of problems.

Parabolic Method

Step 1. Find the equation of the line passing through the end points
of the Pareto optimal line. Use the point-slope formula to obtain
$y = m x + b$.

Step 2. Substitute $y = m x + b$ for y to obtain $(x - x_0)*(mx + b - y_0)$.
As previously shown, the end points selected from our choices are
(3, 4) and (6, 1). The line is $y = -x + 7$, and by substitution, we have
$(x - 1)*(-x + 7 - 1)$. We put into Vertex form to get:

$$y = -1(x - 3.5)^2 + 6.25$$

$x = 3.5$, $Z = 6.25$, and the value of y from $y = -x + 7$ is $y = 3.5$.

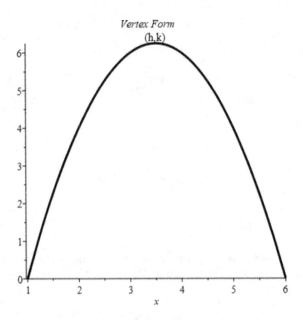

FIGURE 11.3
Parabola method.

Using this method, we found (3.5, 3.5) as the Nash arbitration point.
We find the probabilities identically as before.
This is shown in Figure 11.4.

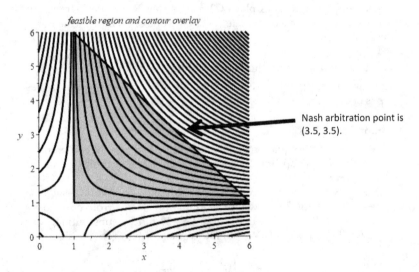

FIGURE 11.4
The feasible region and level curves for $(x - 1)(y - 1)$.

11.3 Nash Arbitration with Calculus

Example 11.2: Nash Arbitration Method

Assume we have the following partial conflict game between two players:

		Player 2	
	Strategies	C1	C2
Player 1	R1	(2, 6)	(5, 4)
	R2	(10, 5)	(4, 8)

Assume in this example that we have already found the Nash equilibrium, (4.666, 5.6). We plot the points and obtain the convex polygon shown in Figure 11.5.

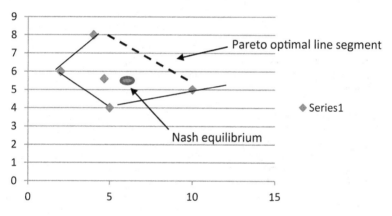

FIGURE 11.5
Payoff polygon and Pareto optimal line segment.

We find the Nash equilibrium is not Pareto optimal. To be Pareto optimal, the Nash equilibrium must lie on the Pareto optimal line segment. We assume that we have tried all other strategic moves' methods to improve our outcomes and now we move on to arbitration. For a good discussion of strategic moves, see Chapter 10.

Finding the Security Levels, (x_0, y_0) by Linear Programming Methods

Two options are given for finding the status quo point, (x_0, y_0). The status quo points can be either the security levels or the threat levels. Every game has security levels and every game may or may not have a threat level. So we teach the use of the security level in this procedure. Security levels are the outcomes of each player's game found using prudential strategies. Games are separated for each player and both players optimize in their own respective games. The method we have taught for finding both the strategies and the game values that work for all matrix games is linear programming, which we have covered in two of our three courses. You can also solve a system of linear equations.

The two games are by each player's perspective:

Player 1's game

	C1	C2
R1	2	5
R2	10	4

Player 2's game

	C1	C2
R1	6	4
R2	5	8

We are more concerned with the solution of each game because those values are our security levels. We use linear programming to find each security level. Our linear programming formulation for each game is as follows:

Maximize SL_I

Subject to:
$$2\,x_1 + 10\,x_2 - SL_I \geq 0$$
$$5\,x_1 + 4\,x_2 - SL_I \geq 0$$
$$x_1 + x_2 = 1$$

Non-negativity

Maximize SL_{II}

Subject to:
$$6\,y_1 + 4\,y_2 - SL_{II} \geq 0$$
$$5\,y_1 + 8\,y_2 - SL_{II} \geq 0$$
$$y_1 + y_2 = 1$$

Non-negativity

The solution of these games yields both the prudential strategies and the security levels that we seek. Player 1 plays 2/3 R1 and 1/3 R2, while Player 2 plays 0.8 C1 and 0.2 C2. The security levels are found as $(x_0, y_0) = (4.6666, 5.6)$.

Alternatively, if linear programming has not been covered in pre-liminary coursework then we can use solving a system of simultaneous equations.

Solving a system of equations consists of three equations and three unknowns; we have for Player 1 the equations:

$$2 x_1 + 10 x_2 - SL_I = 0$$
$$5 x_1 + 4 x_2 - SL_I = 0$$
$$x_1 + x_2 = 1$$

$$\begin{bmatrix} 2 & 10 & -1 & 0 \\ 5 & 4 & -1 & 0 \\ 1 & 1 & 0 & 1 \end{bmatrix}$$

We solve to find the solution 2/3 R1, 1/3 R2, with value 4.6666. We repeat this method for Player 2. The system of equations is:

$$6 y_1 + 4 y_2 - SL_{II} = 0$$
$$5 y_1 + 8 y_2 - SL_{II} = 0$$
$$y_1 + y_2 = 1$$

We solve these systems of equations to obtain 8/10 C1, 2/10 C2, and a value of 5.6.

The security level that we will use in the Nash arbitration scheme is (4.6666, 5.6).

Finding the Pareto Optimal Line Segment Equation

We use the end points (4, 8) and (10, 5) to obtain the equation of the Pareto optimal line shown in Figure 11.4 using the point-slope formula of a line:

$$(y - y_1) = \frac{(y_2 - y_1)}{(x_2 - x_1)}(x - x_1).$$

We find the equation as $(y-8) = -3/6\ (x-4)$, which simplifies to $y = -0.5\ x + 10$.

Nash Arbitration

Now, we can return to the Nash arbitration scheme that says in particu-lar find the values of (x, y) along the line $y = -0.5\ x + 10$ where $x \geq 4.6666$, $y \geq 5.6$ that maximizes the product $(x - 4.6666)\ (y - 5.6)$.

Using Method 1, create a triangle and find the vertices. These vertices are our security level (4.6666, 5.6), (4.6666, 7.6666), and (8.8888, 5.6). The x and y midpoints are 6.733 and 6.633 to four decimal places.

Using Method 2, build the equation of the parabola after substituting into the Nash arbitration equation:

$$(x - 4.6666)*(y - 5.6) \text{ to obtain } (x - 4.666)*(-0.5\ x + 10 - 5.6).$$

This yields the parabola $y = -0.5(x - 6.7333)^2 + 2.136$. We find the x coordinate as 6.7773, and we find the value achieved via arbitration as 2.136. By substituting $x = 6.733$ back into our Pareto optimal line $y = -0.5x + 1$, we find the y coordinate as 6.633.

How is the Arbitration Game Played?

We will need to solve for the probabilities for playing the strategies along the Pareto optimal line segment. We set up simultaneous equations, two equations and two unknowns, to determine what percentage of the time each player plays the end points of the Pareto optimal line. We play point (4, 8) with probability p_1 and (10, 5) with probability p_2 to obtain the solution (6.733241, 6.63338). We solve our system of equations,

$$4\,p_1 + 10\,p_2 = 6.733241$$
$$8\,p_1 + 5\,p_2 = 6.63338$$

obtaining $p_1 = 0.54446$, $p_2 = 0.45554$. Further, we find Player 1 always plays $R2$ so the entire probability is accomplished by Player 2. Player 2 will play $C1$ and $C2$ with probabilities 0.54446 and 0.45554, respectively. We provide the x–y plane view, as displayed in Figure 11.6.

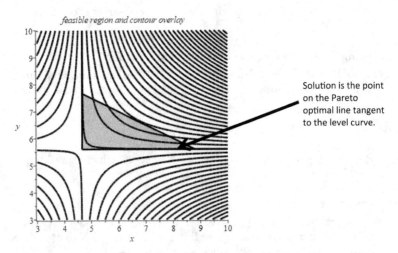

FIGURE 11.6
Nash arbitration and the shaded feasible region of the search.

We also provide a one-dimensional view in Figure 11.7 that shows the search region from 4.6666 to 10 of the parabolic function being searched for the maximum (6.63333, 2.156).

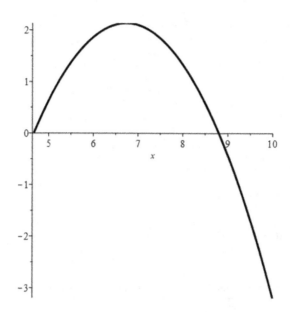

FIGURE 11.7
The function $-0.5\,x^2 + 6.73333\,x - 20.53333$ from $4.6666 \le x \le 10$.

11.4 More than Two Strategies

These methods may be used if the game is played with more than two strategies for each player. We illustrate with an example with three strategies for each player.

Example 11.3: Consider the Following Game

		C1	Colin C2	C3
Rose	R1	(1, 5)	(3, 4)	(2, 0)
	R2	(7, 2)	(3, 3)	(0, 0)

The Nash equilibrium is (6, 1) and the security level is (1, 1). We examine the payoff polygon to visually see the Pareto optimal line so we can determine its equation, see Figure 11.8.

There are two possible Pareto optimal line segments from points (0, 5) to (3, 4) to the north and from (3, 4) to (6, 1) to the east. The method suggested earlier is a little easier. Since we know the security levels (1, 1), we know the function we want to maximize is $f(x, y) = (x-1)(y-1)$. Our corner points of the northeast regions are (0, 5), (3, 4), and (6, 1). We evaluate this function at each corner point and choose the Pareto optimal line endpoints as the two largest.

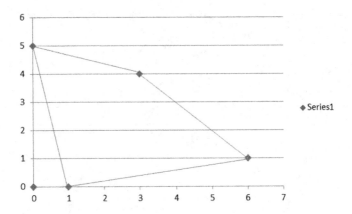

FIGURE 11.8
Payoff polygon.

Corner Point	Evaluation $(x - 1)(y - 1)$
(0, 5)	−4
(3, 4)	6
(6, 1)	0

Points (3, 4) and (6, 1) are the two largest (representing the level curves of the function) so we can use those two points as the end points of the Pareto optimal line segment. The equation is $y = -x + 7$. We find that our Nash arbitration point is (3.5, 3.5) with value $Z = 6.25$. This point (3.5, 3.5) is found by playing 25/30 R1C2 and 5/30 R2C2. We see that Colin always plays C2, so the entire mixed strategy is on Rose. This is shown in Figure 11.9.

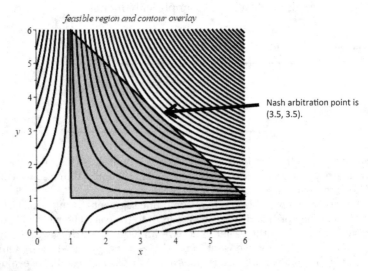

FIGURE 11.9
The feasible region and level curves for $(x - 1)(y - 1)$.

_____

11.5 Writer's Guild Strike Example with Cardinal Numbers

11.5.1 Introduction

In 2007–2008, the Writer's Guild of America went on strike. This strike lasted from November 5, 2007 until February 12, 2008. The strike itself sought to obtain increased funding from the larger studios, the management, huge profit. The 12,000 members of the TV screenwriters and writers went on strike against the Alliance of Motion Picture and Television Producers (*AMPTP*), which was a trade organization that represented the interests of about 397 American film and TV producers see Fox (2015b). According to the analysts,

> At the resolution of the strike, the writers achieved a qualified success: they received a new percentage payment on the distributor's gross for digital distribution, they emerged united, but they lost out on short-term deals see Wikipedia (2007).
>
> WGA members argued that a writer's residuals are a necessary part of a writer's income that is typically relied upon during periods of unemployment common in the writing industry. The WGA requested a doubling of the residual rate for DVD sales, which would result in a residual of 0.6% (up from 0.3%) per DVD sold see Writer's Guild of America (2007).

Let's assume the average TV DVD sells for $10.00. The request for an increase meant that the writers wanted to receive 0.6% instead of 0.3%. This amounts to an increase of $0.03 from $0.03 to $0.06 per sale [1].

The last such strike by the writers in 1988 lasted 21 weeks and 6 days, costing the American entertainment industry an estimated $500 million ($870 million in 2007 dollars) [1]. This strike was just as devastating.

According to a report on the January 13, 2008 edition of NBC Nightly News, if one takes into account everyone affected by the current strike, the strike has cost the industry $1 billion so far; this is a combination of lost wages to cast and crew members of television and film productions and payments for services provided by janitorial services, caterers, prop and costume rental companies, and the like [1].

The TV and movie companies stockpiled "output" so that they could possibly outlast the strike rather than work to meet the demands of the writer's and avoid the strike [1]. From the standpoint of a consumer, the consumers were also not happy.

11.5.2 Game Theory Approach

Let us begin by stating strategies for each player. Our players will be the Writer's Guild and the Management.

11.5.2.1 Strategies

Writer's Guild: Their strategies are to strike (S) or not to strike (NS).

Management: Salary increase plus revenue sharing as the writer's requested (IN), revenue sharing from DVDs only at a rate of 0.6% (GIN), revenue sharing at a substantially smaller rate (RIN), or status quo (SQ).

We will use the alternative method shown by Fox (2015b) in order to create cardinal utilities for the payoff matrix. Another option would be the lottery method by von Neumann and Morgenstern (2004). First, we rank order the outcomes for each side in order of preference. These are the ordinal utilities.

First, we list the combinations of strategies that can be played in this game. This will be a 2×4 game. We also assume that this is a partial conflict (non-zero sum) game. We will refer to NS as R_1, S as R_2, SQ as C_1, In as C_2, RIN as C_3, and Gin as C_4.

$NS\ SQ \rightarrow R_1C_1$

$NS\ IN \rightarrow R_1C_2$

$NS\ GIN \rightarrow R_1C_3$

$NS\ RIN \rightarrow R_1C_4$

$S\ SQ \rightarrow R_2C_1$

$S\ IN \rightarrow R_2C_2$

$S\ RIN \rightarrow R_2C_3$

$S\ GIN \rightarrow R_2C_4$

11.5.2.1.1 Writer's Alternatives and Rankings

The writers have eight strategy combinations to rank. We start by ranking from best to worst. The writers prefer getting compensation to not getting any compensation.

$NS\ IN \rightarrow R_1C_2 = 8$

$NS\ GIN \rightarrow R_1C_3 = 7$

$NS\ RIN \rightarrow R_1C_4 = 6$

$S\ IN \rightarrow R_2C_2 = 5$

$S\ RIN \rightarrow R_2C_3 = 4$

$S\ GIN \rightarrow R_2C_4 = 3$

$NS\ SQ \rightarrow R_1C_1 = 2$

$S\ SQ \rightarrow R_2C_1 = 1$

11.5.2.1.2 *Management's Alternatives and Rankings*

The management prefers not to give additional compensation but if they have to they would prefer at the lower amount. Thus, managements' ordinal ranking might be as follows:

$$NS\ SQ \rightarrow R_1C_1 = 8$$

$$S\ SQ \rightarrow R_2C_1 = 7$$

$$NS\ RIN \rightarrow R_1C_4 = 6$$

$$S\ RIN \rightarrow R_2C_3 = 5$$

$$NS\ GIN \rightarrow R_1C_3 = 4$$

$$S\ GIN \rightarrow R_2C_4 = 3$$

$$NS\ IN \rightarrow R_1C_2 = 2$$

$$S\ IN \rightarrow R_2C_2 = 1$$

Payoff matrix (ordinal)

		Management			
		C1	C2	C3	C4
Writer's	R1	(2, 8)	(8, 2)	(7, 4)	(6, 6)
	R2	(1, 7)	(5, 1)	(4, 5)	(3, 3)

A movement diagram to find a pure strategy equilibrium (see Straffin, 2004). The pure strategy using the ordinal values is R1C1, which represents no strike and status quo. The management is quite happy and the writers are very unhappy. As a matter of fact, the writers begin to consider to strike.

In order to continue the analysis with some accuracy, we need to convert the ordinal values into cardinal values. We could either use the lottery range using the method of von Neumann and Morgenstern (2004) as described by Straffin (2004) or the analytical hierarchy process (AHP) method described by Fox (2015a, 2015c). We illustrate the method described by Fox within this analysis.

We began with analyzing the writer's strategies versus the management's strategies to obtain the ordinal ranking. Now we use Saaty's (1980) nine-point scale to add utility value to those ordinal rankings. We show the pairwise comparison in Table 11.1.

We used an Excel template to input our 1–9 values, develop the matrix, and compute the weights which will be our utility values. These can be viewed in Figure 11.10.

The AHP shows that the matrix, Figure 11.11, that is generated is consistent as the $CR = 0.00073$.

TABLE 11.1

Pairwise Comparison in this Order Using the Nine-Point Scale of Importance

Intensity of Importance in Pairwise Comparisons	Definition
1	Equal importance
3	Moderate importance
5	Strong importance
7	Very strong importance
9	Extreme importance
2,4,6,8	For comparing between the above

Scale for Decision-Criterion and Weights.

The output, the eigenvectors, are the cardinal values for our strategies. These are found for the writers:

R1C2	0.38079158
R1C3	0.22202891
R1C4	0.13584163
R2C2	0.09103997
R2C3	0.0604459
R2C4	0.04445166
R1C1	0.03563703
R2C1	0.02976333

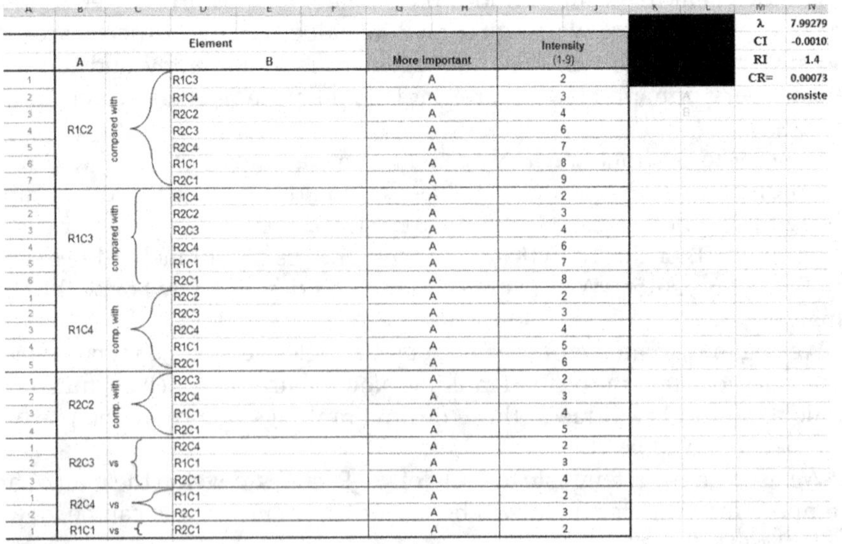

	Element A		Element B	More Important	Intensity (1-9)
1	R1C2	compared with	R1C3	A	2
2			R1C4	A	3
3			R2C2	A	4
4			R2C3	A	6
5			R2C4	A	7
6			R1C1	A	8
7			R2C1	A	9
1	R1C3	compared with	R1C4	A	2
2			R2C2	A	3
3			R2C3	A	4
4			R2C4	A	6
5			R1C1	A	7
6			R2C1	A	8
1	R1C4	comp. with	R2C2	A	2
2			R2C3	A	3
3			R2C4	A	4
4			R1C1	A	5
5			R2C1	A	6
1	R2C2	comp. with	R2C3	A	2
2			R2C4	A	3
3			R1C1	A	4
4			R2C1	A	5
1	R2C3	vs	R2C4	A	2
2			R1C1	A	3
3			R2C1	A	4
1	R2C4	vs	R1C1	A	2
2			R2C1	A	3
1	R1C1	vs	R2C1	A	2

λ	7.99279
CI	-0.0010
RI	1.4
CR=	0.00073
	consiste

FIGURE 11.10
AHP inputs for the Writer's Guild.

Matrix 0

		R1C2	R1C3	R1C4	R2C2	R2C3	R2C4	R1C1	R2C1
		1	2	3	4	5	6	7	8
1	R1C2	1	2	3	4	6	7	8	9
2	R1C3	1/2	1	2	3	4	6	7	8
3	R1C4	1/3	1/2	1	2	3	4	5	6
4	R2C2	1/4	1/3	1/2	1	2	3	4	5
5	R2C3	1/6	1/4	1/3	1/2	1	2	3	4
6	R2C4	1/7	1/6	1/4	1/3	1/2	1	2	3
7	R1C1	1/8	1/7	1/5	1/4	1/3	1/2	1	2
8	R2C1	1/9	1/8	1/6	1/5	1/4	1/3	1/2	1

FIGURE 11.11
Pairwise comparison matrix for writers.

We repeat the identical process for the management. The pairwise comparisons template is shown in Figure 11.12 and the matrix in Figure 11.13.

The output eigenvector, the cardinal values, for our strategies for the management are:

R1C1	0.42608847
R2C1	0.19934334
R1C4	0.11677389
R2C3	0.0807647
R1C3	0.0619435
R2C4	0.04651494
R1C2	0.03758504
R2C2	0.03098613

	Element A		Element B	More Important	Intensity (1-9)
1	R1C1	compared with	R2C1	A	3
2			R1C4	A	4
3			R2C3	A	5
4			R1C3	A	6
5			R2C4	A	7
6			R1C2	A	8
7			R2C2	A	9
1	R2C1	compared with	R1C4	A	3
2			R2C3	A	4
3			R1C3	A	4
4			R2C4	A	6
5			R1C2	A	7
6			R2C2	A	8
1	R1C4	comp. with	R2C3	A	3
2			R1C3	A	3
3			R2C4	A	4
4			R1C2	A	5
5			R2C2	A	6
1	R2C3	comp. with	R1C3	A	3
2			R2C4	A	3
3			R1C2	A	4
4			R2C2	A	5
1	R1C3	vs	R2C4	A	3
2			R1C2	A	3
3			R2C2	A	4
1	R2C4	vs	R1C2	A	3
2			R2C2	A	3
1	R1C2	vs	R2C2	A	3

CI	0.033065
RI	1.4
CR=	0.023618
	consistent

FIGURE 11.12
AHP inputs for the management.

Matrix 0	R1C1	R2C1	R1C4	R2C3	R1C3	R2C4	R1C2	R2C2
	1	2	3	4	5	6	7	8
1 R1C1	1	3	4	5	6	7	8	9
2 R2C1	1/3	1	3	4	4	6	7	8
3 R1C4	1/4	1/3	1	3	3	4	5	6
4 R2C3	1/5	1/4	1/3	1	3	3	4	5
5 R1C3	1/6	1/4	1/3	1/3	1	3	3	4
6 R2C4	1/7	1/6	1/4	1/3	1/3	1	3	3
7 R1C2	1/8	1/7	1/5	1/4	1/3	1/3	1	3
8 R2C2	1/9	1/8	1/6	1/5	1/4	1/3	1/3	1

FIGURE 11.13
Pairwise comparison matrix for management.

This provides us with a payoff matrix consisting of cardinal utilities, Table 11.2. This use of cardinal utilities is important because we really cannot do any mathematics (addition, subtraction, multiplication, or division) with ordinal values. Cardinal values will allow us to compute, as necessary. This will allow us to employ the cardinal values in the Nash arbitration scheme or employ them to see if any mixed strategy Nash equilibrium exists.

Nash proved that every two-person game has at least one equilibrium either in pure or in mixed (equalizing) strategies. The equilibriums are also called Nash equilibriums. He also developed the Nash arbitration method, which was used in this analysis.

11.5.2.2 Finding the Nash Equilibrium

For games with two players and more than two strategies each, we present the nonlinear optimization approach by Barron (2013). Consider a two-person game with a payoff matrix as before. Let's separate the payoff matrix into two matrices **M** and **N** for Players 1 and 2. We solve the

TABLE 11.2

Payoff Matrix for the Game with Cardinal Values

		Management			
		C1	C2	C3	C4
Writers	R1	(0.0356, 0.426)	(0.38079, 0.0375)	(0.222, 0.0619)	(0.1358, 0.1167)
	R2	(0.02976, 0.0309)	(0.091, 0.03098)	(0.0604, 0.0807)	(0.0444, 0.04651)

following nonlinear optimization formulation in expanded form, in Equation (11.1).

$$Maximiz \sum_{i=1}^{n}\sum_{j=1}^{m}x_i a_{ij} y_j + \sum_{i=1}^{n}\sum_{j=1}^{m}x_i b_{ij} y_j + -p - q$$

Subject to:

$$\sum_{j=1}^{m}a_{ij}y_j \leq p,\ i = 1, 2, \ldots, n,$$

$$\sum_{i=1}^{n}x_i b_{ij} \leq q,\ j = 1, 2, \ldots, m, \tag{11.1}$$

$$\sum_{i=1}^{n}x_i = \sum_{j=1}^{m}y_j = 1$$

$$x_i \geq 0, y_j \geq 0$$

We used the computer algebra system Maple to input the game and then solve it. We let the matrix a_{ij} be labeled M and b_{ij} be labeled N in Maple. We wrote a short macro to perform the work.

The commands in Maple are:

- With (LinearAlgerba):with(Optimization)
- A:=Matrix([[0.0356,0.02976],[0.38079,0.091]],[[0.222,0.0619],[0.135,0.0444]]);

$$A := \begin{bmatrix} 0.0356 & 0.02976 \\ 0.38079 & 0.091 \\ 0.222 & 0.0619 \\ 0.1358 & 0.0444 \end{bmatrix}$$

- B:=Matrix([[0.426,0.0309],[0.0375,0.03098]],[0.0619,0.0807],[[0.1167,0.04651]]);

$$B := \begin{bmatrix} 0.426 & 0.0309 \\ 0.0375 & 0.03098 \\ 0.0619 & 0.0807 \\ 0.1167 & 0.04651 \end{bmatrix}$$

- $M:=Transpose(A):N:=Transpose(B):$
- $X:=`<,>`(x[1],x[2]):Y:=`<,>`(y[1],y[2],y[3],y[4]):$
- $c1:=seq(Transpose(X).N([j] \leq q, j=1..4):$
- $c2:=seq(M.Y)[j] \leq p, j=1..2):$

- $c3:=add(x[j],j=1..2)=1:$
- $c4:=add(y[j], j=1..4)=1:$
- $const:=\{c1,c2,c3,c4\};$

$const: = \{x_1 + x_2 = 1, y_1 + y_2 + y_3 + y_4 = 1, 0.0375x_1 + 0.03098x_2 \leq q, 0.0619x_1 +$
$\qquad 0.0807x_2 \leq q, 0.1167x_1 + 0.04651x_2 \leq q, 0.426x_1 +$
$\qquad 0.0309x_2 \leq q, 0.02971y_1 + 0.091y_2 + 0.0619y_3 + 0.0444y_4 \leq p, 0.0356y_1 +$
$\qquad 0.38079y_2 + 0.222y_3 + 0.1358y_4 \leq p\}$

- $objective:=expand(Transpose(X).M.Y + Transpose(X).N.Y-p-q);$

$objective: = 0.4616y_1x_1 + 0.06066y_1x_2 + 0.41829y_2x_1 + 0.12198y_2x_2 +$
$\qquad 0.2839y_3x_1 + 0.1426y_3x_2 + 0.2525y_4x_1 + 0.09091y_4x_2 - p - q$

- $QPSolve(objective, const, assume=nonnegative, maximize, iterationlimit=1000);$

$$\left[2.06519568113350 \; 10^{-9}, \left[p = 0.0355999989674022, q = 0.425999998976, \right. \right.$$
$$\left. \left. y_1 = 1, \; y_2 = 0, y_3 = 0, y_4 = 0, x_1 = 1, x_2 = 0 \right] \right]$$

- QPSolve(objective,   const,   assume=nonnegative,   maximize, intialpoint={p=0.07,q=.59});

$$[0., [p = 0.3560000, q = 0.42600, y_1 = 1, \; y_2 = 0, y_3 = 0, y_4 = 0, x_1 = 1, x_2 = 0]]$$

The Nash equilibrium, as expected is still (0.0356, 0.426) at R1C1. Changing the initial points did not uncover any additional equalizing strategy equilibriums. We also note that this result is not satisfying to the Writer's Guild and that they would like to have a better outcome.

We define the following terms:

Pareto Principle: "To be acceptable as a solution of the game, an outcome should be Pareto Optimal" from Straffin (2004).

Pareto Optimal: The outcome where neither player can improve payoff without hurting (decreasing the payoff) the other player.

As in this case, group rationality (Pareto) is sometimes in conflict with individual rationality (dominant). The eventual outcome depends on the players. Obtaining a Pareto optimal outcome usually requires some sort of communication and cooperation among the players.

With the assumption that the outcome should be Pareto optimal, the next question is, "What is Pareto optimal, and what is it not (Pareto inferior)?" The simplest way for this to be understood is to draw a payoff polygon of the game.

On the chart, the X-axis depicts the payoffs of Rose, and the Y-axis depicts the payoffs of Colin. By plotting the pure strategy solutions on the chart, one can see that the convex (everything inside) polygon enclosing the pure strategy solutions is then the payoff polygon or the feasible region. Therefore, the points inside the polygon are the possible solutions of the game.

Using Excel, we plot these coordinates from the payoff matrix to determine if any points are Pareto optimal. This is the payoff polygon, Figure 11.14.

The Nash equilibrium at (0.0356, 0.426) lies along the Pareto optimal line segment. The writers are still unhappy with the result of this equilibrium. But the writers can do better by going on strike and forcing arbitration. This is exactly what the writers did, they went on strike in order to obtain a better payoff.

We can employ several options to try to secure a better outcome for the writers. We can first try strategic moves and then move to Nash arbitration if the strategic moves did not provide the desired outcomes. Both of these methods employ communications in the game. In strategic moves, we examine the game to see if "moving first" changes the outcome, if threatening our opponent changes the outcome, making promises to our opponent changes our outcome, or a combination of threats and promises changes the outcome.

The result of strategic moves is that (1) moving first, (2) promises, (3) threats, nor (4) combination of threats and promises did not improve the outcome from the Nash equilibrium value of (0.0356, 0.426). The writers have no alternative but to act first and strike to force management's hand. We move on to Nash arbitration.

11.5.2.3 Nash Arbitration Scheme

Nash's theorem states "If SQ(status quo) = (x_0, y_0), then the arbitrated solution point N is the point (x,y) in the polygon with $x \geq x_0$ and $y \geq y_0$ which maximizes the product $(x - x_0) * (y - y_0)$".

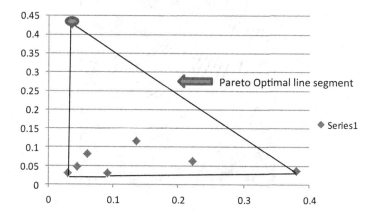

FIGURE 11.14
The payoff polygon.

Status quo point in the definition is the likely outcome of the game when the negotiation fails. An arbitrated solution should be better for both players than the status quo; this is incorporated in the definition by $x \geq x_0$ and $y \geq y_0$. Status quo is the minimum the players can get. Everything above is the improvement of their gain. The solution has to maximize their joint utility. The objective function, $(x - x_0) * (y - y_0)$, maximizes these "above security level" utilities. In other words, it has to maximize the area of the rectangle.

The status quo point can be either the security levels of each side or the threat level. We find these values using prudential strategies. Again, the software can assist us in finding these values as $R1C1$ at $(0.0356, 0.426)$. The security levels do not help the writers so they move on to use the threat level $(0.0297, 0.0309)$. Figure 11.15 shows the feasible region and the level curves $z = (x - 0.0297) * (y - 0.0309)$. From the figure, we can see the approximate Nash arbitration point. We will use nonlinear optimization to find the Nash arbitration point.

11.5.3 Results

We use Maple and find the results. The Nash arbitrated solution is $(0.223, 0.24845)$. The Writer's Guild should be able to improve their outcome from an equilibrium outcome of 0.0356 to 0.223 with an arbitrated solution.

- $f: = (x - 0.02976) - (y - 0.0309)$:
- $constf: = \{y + 1.12547\, x \leq (1.12547 \cdot 0.03580 + 0.46), y > 0.0309\}$:
- $NLPSolve(f, constf, maximize)$;

$$[0.0422061285219754515, [x = 0.223411469519401, y = 0.2488489400]]$$

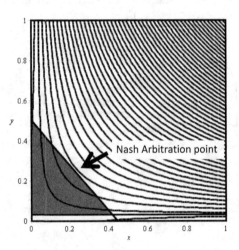

FIGURE 11.15
Nash arbitration plot for the Writer's Guild.

How is this accomplished? In the broader sense, the end points of the Pareto optimal line segment are at the northernmost point most favorable to management and the easternmost point most favorable to the writers. Arbitration is a give-and-take process. In a mathematical sense, the arbitrators should play a mix of (0.0356, 0.426) at *R*1*C*1 and (0.38079, 0.0375) at *R*1*C*2 as these points are the end points of the Pareto optimal line segment. We see clearly that the writers maintain playing *R*1 always and management plays a mix of *C*1 and *C*2 during the process. We find the players should be offing 45.7% *R*1*C*1 and 54.3% *R*2*C*1 to obtain this Nash arbitration point. The Nash arbitration value is the fairest value that the writers can achieve through the strike.

11.5.4 Interpretation

The worst that can happen in the negotiation is to come out at the security level or threat level. For the Writer's Guild, if the threat is indeed successful will most likely move to (0.1358, 0.1167) and this solution is certainly better for the writers than the Nash equilibrium value. The Nash arbitrated solution is (0.2057, 0.24845) and is better for both parties. We found that using the threat level instead of the security levels achieved much better results. This is indeed a better result for the writers, which is what they desired to achieve in a long-term strategy.

The analysis provides insights into the mechanics of the game between the writers and the management. Having a model allows for "what if" analysis to occur in any future negotiations.

We point out that in the short run, the weeks that the writers were on strike, that lost wages and opportunity losses did occur and for management losses also occurred in viewers and sponsorship.

The true results of the strike said that the writers were able to get a new contract that gave them about 0.6% compensation for the DVDs that were sold as stated previously. As a matter of fact, the writers had to go to court to force the management to pay this amount to them.

Chapter 11 Exercises

11.1 Corporation XYZ consists of companies Rose and Colin. Company Rose can make products R1 and R2. Company Colin can make products C1 and C2. These products are not in strict competition with one another, but there is an interactive effect depending on which products are on the market at the same time as reflected in the table below. The table reflects profits in millions of dollars per year.

For example, if products R2 and C1 are produced and marketed simultaneously, Rose's profits are 3 million and Colin's are 6 million annually. Rose can make any mix of R1 and R2, and Colin can make any mix of C1 and C2. Assume the information below is known to each company.

NOTE: The CEO is **not satisfied** with just summing the total profits. He might want the **Nash arbitration point** to award each company proportionately based on their strategic positions if other options fail to produce the results he desires. Further, he does not believe a dollar utility to Rose is of the same importance to the corporation as a dollar utility to Colin.

		COLIN	
		C1	C2
ROSE	R1	(12, 19)	(17,1 4)
	R2	(13, 16)	(14, 17)

1. Suppose the companies have perfect knowledge and implement market strategies independently and simultaneously without communicating with one another. What are the likely outcomes? Justify your choice.

2. Suppose each company has the opportunity to exercise a strategic move. Try only *first moves* for each player and determine if a first move improves the results of the game. (Template available if you need it.)

3. In the event things turn "hostile" between Rose and Colin, find, state, and interpret

   a. Rose's security level and prudential strategy?

   b. Colin's security level and prudential strategy?

   Now assume that the CEO is disappointed with the lack of spontaneous cooperation between Rose and Colin and decides to intervene and wants to dictate the "best" solution for the corporation. He employs an arbiter to determine an "optimal production and marketing schedule" for the corporation.

4. Explain the concept of "Pareto optimal" from the CEO's point of view. Is the "likely outcome" you found in Question 1 at or above Pareto optimal? Briefly explain and provide a payoff polygon plot and draw an arrow to the Pareto optimal line segment.

5. Find and state the Nash arbitration point using the security levels that you found in Question 3.

6. Briefly discuss how you would implement the Nash point. In particular, what mix of the products R1 and R2 should Rose produce

and market, and what mix of the products C1 and C2 should Colin do? Must their efforts be coordinated, or do they simply need to produce the "optimal mix"? Explain briefly.

7. How much annual profit will Rose and Colin each make when the CEO's dictated solution is implemented?

11.2 Corporation XYZ consists of companies Rose and Colin. Company Rose can make products R1 and R2. Company Colin can make products C1 and C2. These products are not in strict competition with one another, but there is an interactive effect depending on which products are on the market at the same time as reflected in the table below. The table reflects profits in millions of dollars per year. For example, if products R2 and C1 are produced and marketed simultaneously, Rose's profits are 3 million and Colin's are 4 million annually. Rose can make any mix of R1 and R2, and Colin can make any mix of C1 and C2. Assume the information below is known to each company.

NOTE: The CEO is **not satisfied** with just summing the total profits. He might want the **Nash arbitration point** to award each company proportionately based on their strategic positions if other options fail to produce the results he desires. Further, he does not believe a dollar to Rose is of the same importance to the corporation as a dollar to Colin.

		COLIN	
		C1	C2
ROSE	R1	(2, 6)	(7, 5)
	R2	(3, 4)	(4, 5)

1. Suppose the companies have perfect knowledge and implement market strategies independently without communicating with one another. What are the likely outcomes? Justify your choice.

2. Suppose each company has the opportunity to exercise a strategic move. Try *first moves* for each player and determine if a first move improves the results of the game.

3. In the event things turn "hostile" between Rose and Colin, find, and then interpret

   a. Rose's security level and prudential strategy?

   b. Colin's security level and prudential strategy?

   Now suppose that the CEO is disappointed with the lack of spontaneous cooperation between Rose and Colin and decides to intervene and dictate the "best" solution for the corporation and employs an arbiter to determine an "optimal production and marketing schedule" for the corporation.

4. Explain the concept of "Pareto optimal" from the CEO's point of view. Is the "likely outcome" you found in Question 1 at or above Pareto optimal? Briefly explain and provide a payoff polygon plot.

5. Find the Nash arbitration point using the security levels found in Question 3.

6. Briefly discuss how you would implement the Nash point. In particular, what mix of the products R1 and R2 should Rose produce and market, and what mix of the products C1 and C2 should Colin do? Must their efforts be coordinated, or do they simply need to produce the "optimal mix"? Explain briefly.

7. How much annual profit will Rose and Colin each make when the CEO's dictated solution is implemented?

References

Barron, E.N. (2013). *Game Theory: An Introduction*. Hoboken, NJ: John Wiley & Sons.

Fox, W.P. (2015a). An alternative approach to the lottery method in utility theory for game theory. *American Journal of Operations Research*. 5 (3), p. 56610. https://doi.org10.4236/ajor.2015.53016.

Fox, W.P. (2015b). Analysis of the 2007-2008 Writer's Guild Strike with game theory, *Applied Mathematics Journal*, 6, pp. 2132–2141.

Fox, W.P. (2015c). The partial conflict game analysis without communication in EXCEL. *Computers in Education Journal*, 6 (4), pp. 2–10.

Nash, J.F. (1950a). Equilibrium points in n-person games. *Proceedings of the National Academy of Sciences of the United States of America*, 36 (1).

Nash, J.F. (1950b). The bargaining problem. *Econometrica*, 18 (2), pp. 155–162.

Saaty, T. (1980). *The Analytical Hierarchy Process*. United States: McGraw Hill.

Straffin, P. (2004). *Game Theory and Strategy*. Washington, DC: Mathematical Association of America, pp. 102–111.

von Neumann, J. & O. Morgenstern. (2004). *Theory of Games and Economic Behavior*, 60th anniversary ed. Princeton, NJ: Princeton University Press.

Additional Readings

Fox, W. (2008). Mathematical modeling of conflict and decision making: The writers' guild strike 2007–2008. *Computers in Education Journal*, 18 (3), pp. 2–11.

Fox, W.P. (2010). Teaching the applications of optimization in game theory's zero-sum and non-zero sum games. *International Journal of Data Analysis Techniques and Strategies (IDATS)*, 2 (3), pp. 258–284.

Fox, W.P. (2012). *Mathematical Modeling with Maple*. Boston, MA: Cengage Publishers, pp. 221–227.

Giordano, F., W. Fox, & S. Horton. (2014). *A First Course in Mathematical Modeling*, 5th ed. Boston, MA: Brooks-Cole.

Giordano, F., W. Fox, & S. Horton. (2014). *A First Course in Mathematical Modeling*. Boston, MA: Cengage Publishers. Chapter 10.

Rao, S. (1979). *Optimization Theory and Application*. New Delhi, India: Wiley Eastern Limited, pp. 205–246.

The New York Times. Who won the writers' strike? *The New York Times*. Retrieved February 12, 2008.

Wikipedia. (2007). The Writer's Guild Strike. http://en.wikipedia.org/wiki/2007_Writers_Guild_of_America_strike.

Winston, W. (1995). *Introduction to Mathematical Programming: Applications and Algorithms*. Belmont, CA: Duxbury Press, pp. 700–709.

Writers Guild of America. (2007). WGA Contract 2007 Proposals (PDF). Retrieved November 9, 2007. https://www.bing.com/ck/a?!&&p=cccdd700d85a87f8Jmlt dHM9MTcyMzY4MDAwMCZpZ3VpZD0yYjg3MzM1NC1mNGE1LTY3MTgtM 2U1ZC0yMTRlZjViNzY2OTgmaW5zaWQ9NTI0MQ&ptn=3&ver=2&hsh=3&fc lid=2b873354-f4a5-6718-3e5d-214ef5b76698&psq=%22WGA+Contract+2007+Pro posals%22+(PDF).+Writers+Guild+of+America.+Archived+(PDF)+from+the+or iginal+on+9+November+2007.&u=a1aHR0cHM6Ly93d3cuaW5keWJlS5vcmcv dXBsb2Fkcy8yMDA3LzEyLzE0L3Byb3Bvc2Fsc2ZbGwyLnBkZg&ntb=1.

12

Three-Person Games

We begin our discussions on n-person games but restrict our analysis to just three-person games.

12.1 Three-Person Zero-Sum Games

Consider the following three-person (total conflict) zero-sum game between Rose, Colin, and Larry each with two strategies (example from Straffin, 2003, Chapter 19).

		Larry D1 Colin	
		C1	C2
Rose	R1	$(1, 1, -2)$	$(-4, 3, 1)$
	R2	$(2, -4, 2)$	$(-5, -5, 10)$

		Larry D2 Colin	
		C1	C2
Rose	R1	$(3, -2, -1)$	$(-6, -6, 12)$
	R2	$(2, 2, -4)$	$(-2, 3, -1)$

We begin our analysis by examining the payoff matrix for pure strategy solutions. We start with the movement diagram as we did with total conflict games in Chapter 5. For Rose and Colin, our movement diagram arrows are the same, but for Larry, the arrows will be placed diagonally in or out from smaller to larger. The movement diagram is displayed in Figure 12.1.

12.1.1 Result from the Movement Diagram

Pure Nash equilibriums at R1C1D2 with payoffs (3, −2, −1) and at R2C1D1 with payoffs (2, −4, 2). These are neither equivalent nor interchangeable. Going for one equilibrium may lead to a nonequilibrium outcome because of

DOI: 10.1201/9781032726885-12

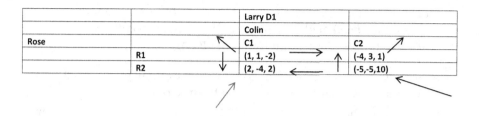

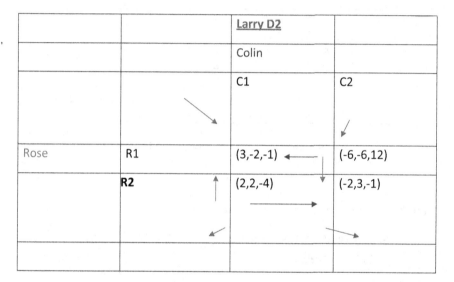

FIGURE 12.1
Movement diagram.

the player's preferences. This leads to a bad situation so perhaps the players will try to form coalitions to obtain a better outcome.

Let's consider communications with the ability to form coalitions. We define a coalition as two players joining against one player.

Assume at first, Rose plays against a coalition of Colin and Larry.

Step 1. Build a payoff matrix for Rose against the Colin-Larry coalition using Rose's values from the original payoffs as follows.

		Colin-Larry			
		C1D1	C2D1	C1D2	C2D2
Rose	R1	1	−4	3	−6
	R2	2	−5	2	−2

Step 2. Look for a solution for the Nash equilibrium using (a) Saddle points (*maximin*) method or (b) Mixed strategies (might need to draw the Rose R1-Rose R2 graph from Chapter 5).

a. No saddle point solution RowMin {−6, −5} ColMax {2, −4, 3, −2}, and the max of Row min does not equal the min of the col Max.

b. The graph shows, Figure 12.2, that the *Maximin* solution is found by using the following values for Rose versus the Coalition. We apply the method of oddments to the remaining two strategies.

		Colin-Larry			
		C2D1	C1D2	Oddments	
Rose	R1	−4	−6	2	3/5
	R2	−5	−2	3	2/5
Oddments		1	4		
		4/5	1/5	Value	−22/5
					−4.4

Note: If the game has a saddle point solution, then that value is the value of the game for all three players. Since we have a mixed strategy, we must find the value for each of our three players.

Step 3. Finding the value of the game for each player.

3/5 3/5 R1C2D1 + 3/5 1/5 R1C2D2 + 2/5 4/5 R2C2D1
+ 2/5 1/5 R2C1D2

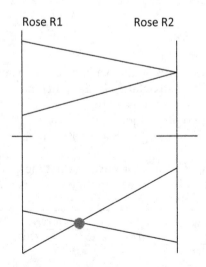

FIGURE 12.2
Graphical analysis leading to a solution.

We now substitute the values from the original payoff matrix.

$$3/5 \ 3/5 \ (-4, 3, 1) + 3/5 \ 1/5 \ (-6, -6, 12) + 2/5 \ 4/5 \ (-5, -5, 10)$$
$$+ 2/5 \ 1/5 \ (-2, 3, -1)$$

Payoff to Rose −4.4, to Colin −0.64, to Larry 5.04.

Step 4. Redo Steps 1–3 for Colin versus a coalition of Rose-Larry and then redo Steps 1–3 for Larry versus a coalition of Rose-Colin.

Colin versus Rose-Larry has a saddle point solution of (2, −4,2). This was the saddle point solution at R2C1D1.

Now, we examine the Rose-Colin coalition.

Larry versus Rose-Colin

		Rose-Colin			
		R1C1	R1C2	R2C1	R2C2
Larry	D1	−2	1	2	10
	D2	−1	12	−4	−1

No saddle point since Max of {−2, −4} is −2 and Min of {−1, 12, 2, 10} is −1.

We now look for a mixed-strategy solution, as shown in Figure 12.3.

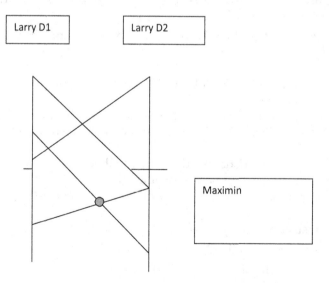

FIGURE 12.3
Graphical method to help find a mixed solution.

Subgame

		Rose-Colin		Oddments	
		R1C1	R2C1	Oddments	
	D1	−2	2	4	3/7
Larry	D2	−1	−4	3	4/7
Oddments		1	6		
		6/7	1/7	Value is	−10/7

$$(3/7)*(6/7)*(1,1,-2)+(3/7)*(1/7)*(2,-4,2)+(4/7)*(6/7)*(3,-2,-1)$$

$$+(4/7)*(1/7)*(2,2,-4)=(104/49,-34/49,\ -10/7)$$

$$=(2.12,-.69,-1.43)\ \text{Rounded to two-decimal places.}$$

Results from these coalition are as follows:

Colin versus Rose-Larry (2, −4, 2) (This was the saddle point solution.)

Larry versus Rose-Colin (2.12, −0.69, −1.43) (see last pages for this reworked).

Rose versus Colin-Larry (−4.4, −0.64, 5.04) (from before).

Step 5. Determine which coalition yields the best payoff for each player.

Rose: Max {2, 2.12, −4.4} is 2.12 so Rose prefers a coalition with Colin.

Colin: Max {−4, −0.69, −0.64} is −0.64 so Colin prefers a coalition with Larry.

Larry: Max {2, −1.43, 5.04} is 5.04 so Larry prefers a coalition with Colin.

In two of these cases, Colin-Larry is the preferred coalition so we might expect that coalition to be formed.

Note: we may or may not be able to determine which coalition might be formed. Note on SIDEPAYMENTS: these are bribes or payments to entice a coalition.

Note of the characteristic function: The number $v(S)$, called the value of S, is to be interpreted as the amount S would win if they formed a coalition. We assume that the empty coalition (none are formed) value is 0, $v(\varnothing) = 0$

Colin versus Rose-Larry (2, −4, 2)

Larry versus Rose-Colin (2.12, −0.69, −1.43)

Rose versus Colin-Larry (−4.4, −0.64, 5.04)

We can build the functions used in Shapely Analysis next.

12.1.2 Shapely Analysis

Empty set: $v(\emptyset) = 0$

$$\text{Alone: v(Rose)} = -4.4, \text{v(Colin)} = -4, \text{v(Larry)} = -1.43$$

Coalition by twos:

$$\text{v(Rose-Colin)} = 1.43 \; \text{v(Rose-Larry)} = 4 \; \text{v} = \text{(Colin-Larry)} = 4.4$$

We add the payoff for the coalition's partners in the associated games. Coalitions by three: These are zero-sum games so adding all payoff together = 0 v(Rose-Colin-Larry) = 0

12.2 Three-Person Partial Conflict Game (Non-Zero-Sum Game)

12.2.1 Constant Sum

Assume we have another three-person game that is a non-zero sum but a constant sum game since all sums equal 10.

The movement diagram gives a pure strategy solution at R1C1D1, (4, 3, 3).

		Larry D1 Colin			Larry D2	
		C1	C2		C1	C2
Rose	R1	(4, 3, 3)	(1, 2, 7)	R1	(3, 6, 1)	(2, 5, 3)
	R2	(3, 5, 2)	(0, 4, 6)	R2	(2, 7, 1)	(1, 6, 3)

We set up and solve all the possible coalitions: Rose versus Colin-Larry

		Colin-Larry			
		C1D1	C2 D1	C1D2	C2D2
Rose	R1	4	1	3	2
	R2	3	0	2	1

Solution is a Pure Strategy at R1C2D1 at (1, 2, 7) through a saddle point. Again, looking for a saddle point in a coalition game is still a technique to try first.

We could have obtained this answer via linear programming (LP) as well.

Max v St.

$$4x1 + 3x2 - V \geq 0$$
$$x1 - V \geq 0$$
$$3x1 + 2x2 - V \geq 0$$
$$2x1 + x2 - V \geq 0$$
$$x1 + x2 = 1x1 \leq 1, x2 \leq 1$$

The solution for Rose has $x1 = 1$, $x2 = 0$, Value = 1 that implies pure strategy solution shown in Figure 12.4.

$$y1 = 0, \ y2 = 1, \ y3 = 0, \ y4 = 0 \text{ leading to R1C2D1 at } (1, 27).$$

Every coalition in this example leads to pure strategy solution via the saddle point solution method. If Rose and Larry form a coalition against Colin, we find a saddle at (4, 3, 3) the Nash Equilibrium. If Rose and Colin form a coalition against Larry, we find a saddle at (3, 5, 2).

Colin versus Rose-Larry

	R1D1	R2 D1	R1D2	R2D2
Colin	3	5	6	7
C1				
C2	2	4	5	6

Saddle at   (4, 3, 3)

Microsoft Excel 11.0 Sensitivity Report
Worksheet: [test n person game.xls]Sheet2
Report Created: 2/23/2011 2:52:40 PM
Adjustable Cells

Cell	Name	Final Value	Reduced Cost	Objective Coefficient	Allowable Increase	Allowable Decrease
$J$3	x1	1	0	0	1E+30	1
$J$4	x2	0	−1	0	1	1E+30
$J$5	V	1	0	1	1E+30	1

Constraints

Cell	Name	Final Value	Shadow Price	Constraint R.H. Side	Allowable Increase	Allowable Decrease
$L$8	used	3	0	0	3	1E+30
$L$9	used	0	−1	0	1	1
$L$10	limit used	2	0	0	2	1E+30
$L$11	used	1	0	0	1	1E+30
$L$12	used	1	1	1	0	1
$L$13	used	1	0	1	1E+30	0
$L$14	used	0	0	1	1E+30	1

FIGURE 12.4
Screenshot of Linear Programming (LP) solution from EXCEL.

Larry versus Colin-Rose

	R1C1	R2 C1	R1 C2	R2 C2
Larry D1	3	2	7	6
D2	1	1	3	3

Saddle at $(3, 5, 2)$

Interpretation: Colin would not join a coalition with Larry because it gives him a lower payoff than the others' choices. We must consider side payments (bribes) as a possibility.

12.3 A Three-Person Game that is Strictly a Non-Zero Non-Constant Sum Game

Start with the movement diagram shown in Figure 12.5. The movement diagram gives R2C2D2 at $(-1, -1, -1)$ as the Nash equilibrium. This is Pareto inferior. We would all like $(1, 1, 1)$ at R1C1D1 better. Thus, operating alone is not so good.

Consider the coalitions: Colin and Larry against Rose

Rose	C1D1	C2D1	C1D2	C2D2
R1	(1, 2)	(0, 3)	(0, 3)	(−2, 4)
R2	(3, 0)	(2, 0)	(2, 0)	(−1, −2)

Rose's game:

Rose	C1D1	C2D1	C1D2	C2D2	Rowmin	Max
R1	1	0	0	−2	−2	
R2	3	2	2	−1	−1	−1
ColMax	3	2	3	−2		
Min				−2		

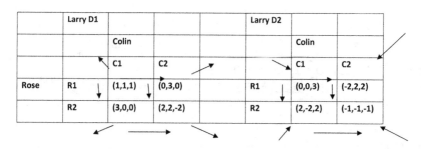

FIGURE 12.5
Movement diagram.

Saddle point at (−1, −1, −1) which is also Rose's security level {−1}. Still not good.

Colin's game:

Colin	R1D1	R2D1	R1D2	R2D2	Rowmin
Max C1	2	0	3	0	0
	0				
C2	3	0	4	−2	−2
ColMax	3	0	4	0	
Min		0		0	Tie

Saddle point solutions at R2C1D1 (2,0) AND R2C1D2 AT (2,0), SO Colin and Larry each get 0. Each coalition ends up the same, with the coalition getting 0 as their game value.

We must find a way to convince all players to choose R1, C1, D1.

If all players agree to play R1C1D1 then they each get (1, 1, 1) – the best result.

12.4 Non-Zero Sum (Non-Constant Sum) with No Pure Strategies

Creating a new game that has no pure strategy Nash equilibriums as the movement diagram, Figure 12.6, is revolving:

The movement diagram, Figure 12.6, yields no pure strategy Nash equilibrium.

Let's examine the coalitions starting with Rose versus Colin-Larry.

	C1D1	C2D1	C1D2	C2D2
R1	(2, 3)	(4, 3)	(4, 5)	(5, 3)
R2	(3, 2)	(3, 1)	(5, 3)	(3, 3)

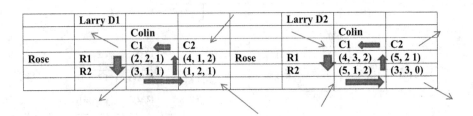

FIGURE 12.6
Movement diagram.

Solving for what Rose should do.

	C1D1	C2D1	C1D2	C2D2	Rowmin
R1	2	4	4	5	2
R2	3	3	5	3	**3**
colMax	**3**	4	5	5	

At saddle point (3, 2), which is (3, 1, 1), what should the coalition do?

	R1	R2	
C1D1	3	2	2
C2D1	3	1	1
C1D2	5	3	**3**
C2D2	3	3	**3**
	5	**3**	

Saddle at R2C2D2 (3, 3, 0) – if the coalition shares equally – (3, 1.5, 1.5) is preferable at this point.
Coalition Rose versus Colin-Larry

		Larry D1			**Larry D2**	
		Colin			**Colin**	
		C1	C2		C1	C2
Rose	R1	(2, 2, 1)	(4, 1, 2)	R1	(4, 3, 2)	(5, 2, 1)
	R2	(3, 1, 1)	(1, 2, 1)	R2	(5, 1, 2)	(3, 3, 0)

Examine the coalition Colin versus Rose-Larry

	R1D1	R2D1	R1D2	R2D2
C1	(2, 3)	(1, 4)	(3, 6)	(1, 7)
C2	(1, 6)	(2, 2)	(2, 6)	(3, 3)

Solving for what Colin should do.

	R1D1	R2D1	R1D2	R2D2	Rowmin
C1	2	1	3	1	1
C2	1	2	2	3	1
colMax	2	2	3	3	**No saddle**

Using Williams' graphical methods, Chapter 5, we find by playing 2/9
R1C1D2, 1/9 R2C2D2, 4/9 R1C2D1, 2/9 R2C2D2 we get a value of 7/3 for
Colin. If this happens, the coalition gets to share: (2/9) 6 + (1/9) 6 + (4/9) 6 +
(2/9) (3) = 48/9 = 5 4/9.
If the coalition optimizes then we have

	C1	C2	Row min
R1D1	3	6	3
R2D1	4	2	2
R1D2	6	6	6
R2D2	7	3	3
Col Max	7	6	Saddle at 6

Note that 6 is preferable to 5 and 4/9. Coalition forces R1C2D2 at (4, 3, 2)
which is worse for Colin and better for Rose-Larry.
What should the coalition do?

	R1	R2	
C1D1	3	2	2
C2D1	3	1	1
C1D2	5	3	3
C2D2	3	3	3
	5	3	

Saddle at R2C2D2 (3, 3, 0) – if the coalition shares equally – (3, 1.5, 1.5) is
preferable at this point.

Coalition Colin versus Rose-Larry

Last coalition: Larry versus Rose-Colin

Coalition Colin versus Rose-Larry

		Larry D1			Larry D2	
		Colin			Colin	
		C1	C2		C1	C2
Rose	R1	(2, 2, 1)	(4, 1, 2)	R1	(4, 3, 2)	(5, 2, 1)
	R2	(3, 1, 1)	(1, 2, 1)	R2	(5, 1, 2)	(3, 3, 0)

Examining the coalition of Larry versus Rose-Colin

	R1C1	R2C1	R1C2	R2C2
D1	(1, 4)	(1, 4)	(2, 5)	(1, 3)
D2	(2, 7)	(2, 6)	(1, 7)	(0, 6)

Solving for what Larry should do.

	R1C1	R2C1	R1C2	R2C2	Rowmin
D1	1	1	2	1	1
D2	2	2	1	0	0
colMax	2	2	2	1	**Saddle**

Saddle found at R2C2D1 (1, 2, 1) so Larry gets 1 and the coalition gets 3 (1 for Rose and 2 for Colin or a 1.5 split).

If the coalition optimizes then we have

	D1	D2	Row min
R1C1	4	7	4
R2C1	4	6	4
R1C2	5	7	5
R2C2	3	6	3
Col Max	**5**	**7**	Saddle at 5

Saddle at R1C2D1 (4, 1, 2). This is better for all parties.

Ultimately greed prevails and whichever combination produces the higher values to the players (even with bribes) will most likely work.

12.5 Three-Person Game with Technology

Let's revisit our first example from this chapter and use technology. For our course, we developed an EXCEL template to assist in solving three-person games. Although the mathematics is not difficult, the number of calculations is quite tedious. Therefore, we built a technology assistant for student use.

12.5.1 Technology Assistant with EXCEL

We developed a technology assistant to assist the students with the many calculations involved. Instructions are provided within the template, which is a macro-enhanced EXCEL worksheet. These instructions include:

1. **Put the R,C,L entries into the blocks to the left.**
2. **Go to Coalition_R_CL and execute the Solver.**
3. **Go to Coalition_C_RL and execute the Solver.**
4. **Go to Coalition_L_RC and execute the Solver.**

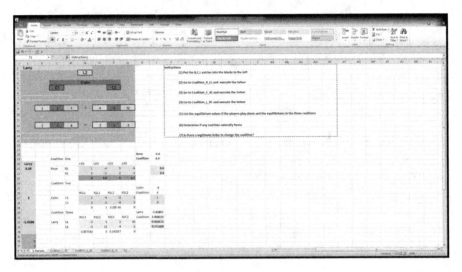

FIGURE 12.7
Screen shot of three-person game template with instructions.

5. **List the equilibrium values if the players play alone and the equilibriums in the three coalitions.**

6. **Determine if any coalition naturally forms.**

7. **Is there a legitimate bribe to change the coalition?**

In Figure 12.7, we find the results or outcomes of the calculations made to find the pure strategies equilibrium and the results of the coalitions. The user must then interpret the results and make conclusion about those results as to what is likely to occur.

12.5.2 N-Person Games with Linear Programming

The coalition's solution on each worksheet uses the Solver, specifically SimplexLP. We illustrate this with a three-person zero-sum game that we just saw in a previous example. Recall that we created the game payoffs for the potential coalitions:

		Colin-Larry			
		C1L1	C2L1	C1L2	C2L2
Rose	R1	1	−4	3	−6
	R2	2	−5	2	−2

This is a zero-sum game for solve for Rose and get the Colin-Larry coalition's results come from the sensitivity column. Note that there are some negative entries as payoffs so we let $v = V1 - V2$ (see Winston, 1995). We formulate the LP.

Maximize $v = V1 - V2$
$x1 + 2x1 - V1 + V2 \geq 0$
$-4x1 - 5x2 - V1 + V2 \geq 0$
$3x1 + 2x2 - V1 + V2 \geq 0$
$-6x1 - 2x2 - V1 + V2 \geq 0$
$x1 + x2 = 1$
$x1 \leq 1$
$x2 \leq 1$
non-negativity

We find the LP solution to this game for Rose is -4.4 when $x_1 = 0.6$ and $x_2 = 0.4$. We find from the reduced costs (the dual solution for Colin and Larry coalition),

$$y1 = y3 = 0, y2 = 0.8 \text{ and } y4 = 0.2, Vcl = 4.4$$

Although this gives us a coalition value, we must use all the probabilities of players to obtain the values for each of our player separately. We only have to use the strategies with probabilities greater than 0:

$(.6)(.8) R1C2L1 + (.4)(.8) R2C2L1 + (.6)(.2) R1C2L2 + (.4)(.2) R2C2L2$

$.48(-4, 3, 1) + .32 (-5, -5, 10) + .12(-6, -6, 12) + .08(-2, 3, -1) = (-4.4, -0.64, 5.04)$

Rose loses -4.4 (as shown before) and the coalition 4.4 is broken down as -0.64 for Colin and 5.04 for Larry.

We repeat this process for each Coalition to obtain these results:

Colin versus Rose-Larry $(2, -4, 2)$
Larry versus Rose-Colin $(2.12, -0.69, -1.43)$

It is still up to the user to interpret and analyze these results These procedures work for constant sum games as well.

12.5.3 A Three-Person Game that is a Strict Non-Zero-Sum Game Using Technology

We also developed an assistant for the partial conflict game. This technology assistant requires the use of the Solver six times in the spreadsheet since each player or side in a coalition requires a LP solution. The instructions are listed inside the template, shown in Figure 12.8.

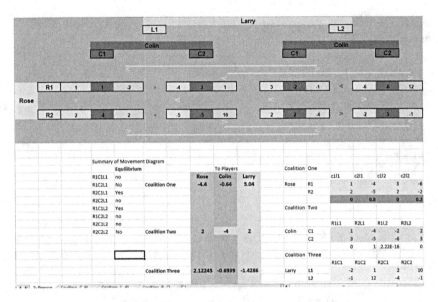

FIGURE 12.8
Screenshot.

The results here are as follows:

Pure strategy by movement
diagram finds an equilibrium at
R1C1L2 with values (2, 1, 1)

	Equilibrium
R1C1L1	No
R1C2L1	No
R2C1L1	No
R2C2L1	No
R1C1L2	Yes
R1C2L2	No
R2C1L2	No
R2C2L2	No

We easily see a better set of values as an output of (4, 2, 3) at R1C2L1. We analyze all coalitions to see if that solution rises from any coalitions.

From the LP solution of the coalitions, we find solutions shown in Figure 12.9:

Rose prefers a coalition with Larry, Colin prefers a coalition with Rose, and Larry prefers either a coalition with Colin or being alone. There is no preferred coalition and none gets us to the better value.

Perhaps all the players should just all agree to play the strategies that provide the best solution.

	To Players		
	Rose	**Colin**	**Larry**
Coalition One Rose VS Colin- Larry	1.5	1	1
Coalition Two Colin vs Rose- Larry	1.75	0.5	0.75
Coalition Three Larry vs Rose- Colin	1.5	1.5	1

FIGURE 12.9
Screenshot of solution.

12.5.4 Conclusions

We have described the use of EXCEL templates to assist in the solution to the three-person games. We remark that users must still analyze the numerical values to determine what will most likely happen. The author will provide these templates upon request. Email requests to wpfox@nps.edu.

Chapter 12 Exercises

Solve the following three-person games.

12.1 Consider the following three-person zero-sum game between Rose, Colin, and Larry

		Larry D1	
		Colin	
		C1	**C2**
Rose	R1	(5, 5, −10)	(−2, 3, −1)
	R2	(4, −5, 1)	(3, −2, −1)

		Larry D2 Colin	
		C1	**C2**
Rose	R1	(−2, 4, −,2)	(−1, −2, 3)
	R2	(−4, 5, −1)	(1, 2, −3)

12.2 Consider the following three-person zero-sum game between Rose, Colin, and Larry

		Larry D1 Colin	
		C1	**C2**
Rose	R1	(5, 5, 10)	(2, 3, 1)
	R2	(4, 5, 1)	(3, 2, 1)

		Larry D2 Colin	
		C1	**C2**
Rose	R1	(2, 4 2)	(1, 2, 3)
	R2	(4, 5, 1)	(1, 2, 3)

12.3 Consider the following three-person zero-sum game between Rose, Colin, and Larry

		Larry D1 Colin	
		C1	**C2**
Rose	R1	(70, 70, 70)	(10, 10, 23)
	R2	(60, 0, 0)	(65, 65 10)

		Larry D2 Colin	
		C1	**C2**
Rose	R1	(70, 70, 60)	(10, 20, 0)
	R2	(80, 50, 30)	(60, 55, 5)

References

Straffin, P. (2003). *Game Theory and Strategy*. Washington, DC: The Mathematical Association of America. Chapter 19.

Winston, W. (1995). *Introduction to Mathematical Programming: Applications and Algorithms*. Belmont, CA: Duxbury Press.

Additional Readings

Fox, W. (2012). *Mathematical Modeling with Maple*. Boston, MA: Cengage Publishers, pp. 221–227.

Fox, W.P. (2010). Teaching the applications of optimization in game theory's zero-sum and non-zero sum games. *International Journal of Data Analysis Techniques and Strategies (IDATS)*, 2 (3), pp. 258–284.

Fox, W.P. (2015). The partial conflict game analysis without communication in EXCEL. *Computers in Education Journal*, 6 (4), pp. 2–10.

13

Extensive-Form Games

13.1 Introduction to Extensive Games

What is an extensive-form game?

From Wiki

In game theory, an **extensive-form game** is a specification of a game allow-ing (as the name suggests) for the explicit representation of a number of key aspects, like the sequencing of players' possible moves, their choices at every decision point, the (possibly imperfect) information each player has about the other player's moves when they make a decision, and their payoffs for all possible game outcomes. Extensive-form games also allow for the repre-sentation of incomplete information in the form of chance events modeled as "moves by nature". Extensive-form representations differ from normal-form in that they provide a more complete description of the game in question, whereas normal-form simply boils down the game into a payoff matrix.

Thus, instead of simultaneous actions, we find we may model actions sequentially.

Examples might include the game of 21, poker, and craps where the player gets to see an event before making a subsequent decision.

In general, we have two players: Player 1 and Player 2, and assume each has two decisions: Run or Hide. The payoffs are seen in the tree diagram

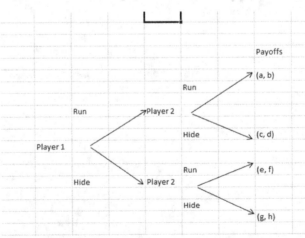

DOI: 10.1201/9781032726885-13

We use backward induction to solve these extensive games. We start at the payoffs and work backward. Under Player 1 Run, Player 2 has two options Run (a, b) and Hide (c, d). If d > b, then Player 2 chooses (c, d), and if b > d, then Player 2 chooses (a, b). If Player 1 chooses Hide, then Player 2 compares Run at (e, f) to Hide at (g, h). If h > f, then Player 2 chooses (g, h), otherwise Player 2 chooses (e, f).

Assume d > b and h> f, so Player 1 can Run and compare (c, d) to Hide at (g, h). If c > g, then Player 1 chooses Run but if g > c then Player 1 chooses Hide.

We illustrate a few of the more straightforward sequential games.

Example 13.1: Kidnapping for Ransom

We will begin with a kidnapping scenario as seen in several TV episodes. In this example, a kidnapper takes a hostage and demands that the hostage pays a ransom. The hostage may pay the ransom or not pay the ransom. Following that the kidnapper may kill the hostage or may not kill the hostage and release the hostage. If the hostage is released then he may or may not report the kidnapping to the police.

We make up some numbers to illustrate the scenario. Assume the kidnapper get +5 for getting paid, −2 for having the kidnapping reported to the police, and −1 for killing the hostage. We assume that these utilities are additive. The hostage's utilities are −10 for being killed, −2 for paying the kidnapper, and +1 for being released and reporting the kidnapping. Entries will be (kidnapper, hostage). For example, pay, release, and report would be calculated as (+5 −2) = 3 and −2 +1 = −1) = (3, −1). The game tree is shown in Figure 13.1.

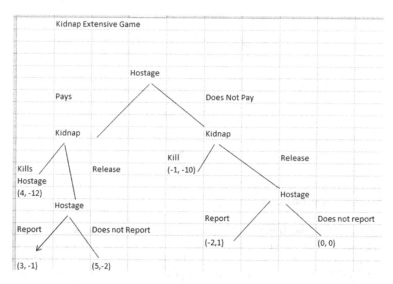

FIGURE 13.1
Kidnapping game tree.

Solution: We use backward induction. We start at the bottom (or end of the tree) and work up or back to the beginning of the tree. We bring the best outcome at each stage. For the hostage, along the pay branch, we compare report the kidnapping (3, −1) to do not report (5, −2). I is better to report the kidnapping at values (3, −1). Next, we compare for the kidnapper the result of kill the hostage (4, −12) to the value we just brought up (3, −1). The kidnapper prefer (4, −12) over (3, −1). Next, we examine the don't pay branch. First, for the hostage, we compare report the kidnapping at (−2, 1) to do not report at (0, 0). The hostage prefers to report at (−2, 1) over (0, 0). We move up the tree and now compare the values at kill (−1, −10) to the value we just brought forward (−2, 1). The kidnapper prefer (−1, −10). Now the hostage compares the result of pay (4, −12) to do not pay at (−1, −10). The hostage prefers do not pay at (−1, −10). The Nash equilibrium is (−1, 10) where the hostage does not pay and the kidnapper kills the hostage.

Discussion: This is why in most TV scenarios, the police are doing everything not to prevent this outcome.

Example 13.2: North Korea-United States Game

First, we consider a scenario about North Korea and their missile tests and closest to a real event. We provide values from Professor Loentz during a presentation at the Naval Postgraduate School in 2015. The game tree is shown in Figure 13.2.

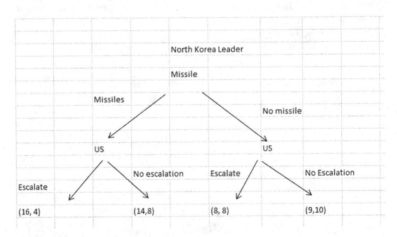

FIGURE 13.2
North Korea missile game tree.

Solution: Under the branch of missiles, the United States is better off not escalating with (14, 8) better than (16,4), and under the no missiles branch, we find no escalation (9, 10) is better than escalation (8, 8). Under the North Korean decision to use missiles or not use missile, we find (14, 8) is better than (9, 10). So we find that North Korea is better off having the missiles and for the United States not to escalate.

Example 13.3: North Korean-US Missiles Revisited

This is a scenario about North Korea and its leader Kim Jong Un. First, we consider his actions. We assume that 25% of the time his decisions are crazy and 74% of the time they are not crazy. We know they are testing missiles. He can fire his test missile or not fire a test missile. The United States can then either escalate or not escalate. This is shown in Figure 13.3.

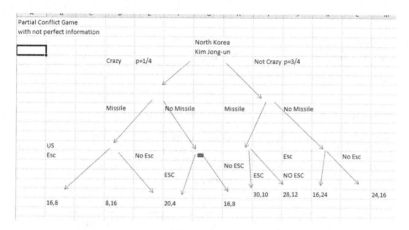

FIGURE 13.3
North Korea-United States with a twist.

We might work backward with induction or collapse this scenario into a matrix game.

First, compute pairwise strategies calculations based upon expected value.

Calculations	
ESC-Missile	$1/4\,(16, 8) + (3/4)\,(30, 10)$
ESC NO Missile	$1/4\,(20, 4) + (3/4)\,(16, 24)$
No ESC-Missile	$1/4\,(8, 16) + (3/4)\,(28, 12)$
No ESC-No Missile	$1/4\,(16, 8) + (3/4)\,(24, 16)$

We might even collapse this into a payoff matrix, as shown in Figure 13.4. The decision from this payoff matrix is no missiles by North Korea and no escalation by the United States using movement diagrams.

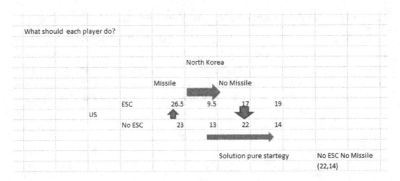

FIGURE 13.4
Payoff matrix for North Korean-US missile.

Example 13.4: Cuban Missile Crisis as an Extensive Game

Although we examined this as a game of chicken and with communications using strategic moves, we will re-examine here as a sequential game. In this scenario, we have the USSR leader, Khrushchev, placing missiles in Cuba and the US President Kennedy reacting.

		Kennedy		
		Do nothing	Blockade	Air Strike
Khrushchev	Don't place missiles in Cuba	u	u	u
	Place but always acquiesce	v	w	y
	Place always escalate	v	x	z
	Place escalation only with blockade	v	x	y
	Place only escalate with air strike	v	w	z

We begin with ordinal values for the values.
This is a partial conflict game.

		Kennedy		
		Do nothing	Blockade	Air Strike
Khrushchev	Don't place missiles in Cuba	(5, 4)	(5, 4)	(5, 4)
	Place but always acquiesce	(6, 3)	(4, 6)	(3, 5)
	Place always escalate	(6, 3)	(1, 2)	(2, 1)
	Place escalation only with blockade	(6, 3)	(1, 2)	(3, 5)
	Place only escalate with air strike	(6, 3)	(4, 6)	(2, 1)

There are three Nash equilibriums: "Don't place missile, Blockade" that corresponds to the backward induction strategies from the tree, "Don't place missile, Air Strike" results in the same outcome values, and "Place missile and always escalate, Do Nothing" which involves an incredible threat (as presented in strategic moves) in order to keep Kennedy from doing something. Perhaps this explains why the USSR was so secretive about the missile placement in the first place (see Figure 13.5).

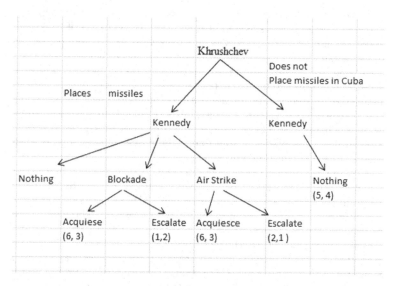

FIGURE 13.5
USSR-United States in Cuban Missile Crisis.

Applying the Theory of Moves to alter the outcomes,
To determine where play will end up when at least one player wants to move from the initial state, I assume the players use *backward induction*. This is a reasoning process by which the players, working backward from the last possible move in a game, anticipate each other's rational choices. For this purpose, I assume that each has complete information about the other's preferences, so each can calculate the other player's rational choices, as well as its own, in deciding whether to move from the initial state or any subsequent state.
To illustrate backward induction, consider again the game Alternative in Figure 13.2. After the missiles were detected and the United States imposed a blockade on Cuba, the game was in state Blockade Maintain Missile (BM), which is worst for the United States (1) and best for the Soviet Union (4). Now consider the clockwise progression of moves that the United States can initiate by moving to Air Strike and Maintain Missiles (AM), the Soviet Union to Air Strike and Withdraw Missiles (AW), and so on, assuming the players look ahead

to the possibility that the game makes one complete cycle and returns to the initial state (State 1):

	State 1		State 2		State 3		State 4		State 1
U.S.	U.S.	→	S.U.	→	U.S.	→|	S.U.	→	(1, 4)
starts	(1, 4)		(4, 1)		(2, 2)		(3, 3)		
Survivor	(2, 2)		(2, 2)		(2, 2)		(1, 4)		

This is a game tree, though drawn horizontally rather than vertically. The survivor is a state selected at each stage as the result of backward induction. It is determined by working backward from where play, theoretically, can end up (State 1, at the completion of the cycle).

Assume the players' alternating moves have taken them clockwise in Alternative from (1, 4) to (4, 1) to (2, 2) to (3, 3), at which point S.U. in State 4 must decide whether to stop at (3, 3) or complete the cycle by returning to (1, 4). Clearly, S.U. prefers (1, 4) to (3, 3), so (1, 4) is listed as the survivor below (3, 3): because S.U. would move the process back to (1, 4) should it reach (3, 3), the players know that if the move-countermove process reaches this state, the outcome will be (1, 4).

Knowing this, would U.S. at the prior state, (2, 2), move to (3, 3)? Because U.S. prefers (2, 2) to the survivor at (3, 3) – namely, (1, 4) – the answer is no. Hence, (2, 2) becomes the survivor when U.S. must choose between stopping at (2, 2) and moving to (3, 3) – which, as I just showed, would become (1, 4) once (3, 3) is reached.

At the prior state, (4, 1), S.U. would prefer moving to (2, 2) than stopping at (4, 1), so (2, 2) again is the survivor if the process reaches (4, 1). Similarly, at the initial state, (1, 4), because U.S. prefers the previous survivor, (2, 2), to (1, 4), (2, 2) is the survivor at this state as well.

The fact that (2, 2) is the survivor at the initial state, (1, 4), means that it is rational for U.S. to move to (4, 1), and S.U. subsequently to (2, 2), where the process will stop, making (2, 2) the rational choice if U.S. moves first from the initial state, (1, 4). That is, after working backward from S.U.'s choice of completing the cycle or not from (3, 3), the players can reverse the process and, looking forward, determine what is rational for each to do. I indicate that it is rational for the process to stop at (2, 2) by the vertical line blocking the arrow emanating from (2, 2), and underscoring (2, 2) at this point.

Observe that (2, 2) at state AM is worse for both players than (3, 3) at state BW. Can S.U., instead of letting U.S. initiate the move-countermove process at (1, 4), do better by seizing the initiative and moving, counterclockwise, from its best state of (1, 4)? Not only is the answer yes, but it is also in the interest of U.S. to allow S.U. to start the process, as seen in the following counterclockwise progression of moves from (1, 4):

	State 1		State 2		State 3		State 4		State 1
S.U.	S.U.	→	U.S.	→|	S.U.	→	U.S.	→	(1, 4)
starts	(1, 4)		(3, 3)		(2, 2)		(4, 1)		
Survivor	(3, 3)		(3, 3)		(2, 2)		(4, 1)		

S.U., by acting "magnanimously" in moving from victory (4) at BM to compromise (3) at BW, makes it rational for U.S. to terminate play at (3, 3), as seen by the blocked arrow emanating from State 2. This, of course, is exactly what happened in the crisis, with the threat of further escalation by the United States, including the forced surfacing of Soviet submarines as well as an air strike (the U.S. Air Force estimated it had a 90% chance of eliminating all the missiles), being the incentive for the Soviets to withdraw their missiles.

Applying Theory of Moves (TOM)

Like any scientific theory, the theory of moves (TOM) calculations may not take into account the empirical realities of a situation. In the second backward-induction calculation, for example, it is hard to imagine a move by the Soviet Union from States 3 to 4, involving maintenance (via reinstallation?) of their missiles after their withdrawal and an air strike. However, if a move to State 4, and later back to State 1, were ruled out as infeasible, the result would be the same: commencing the backward induction at State 3, it would be rational for the Soviet Union to move initially to State 2 (compromise), where the play would stop.

Compromise would also be rational in the first backward-induction calculation if the same move (a return to maintenance), which in this progression is from State 4 back to State 1, were believed infeasible: commencing the backward induction at State 4, it would be rational for the United States to escalate to air strike to induce moves that carry the players to compromise at State 4. Because it is less costly for both sides if the Soviet Union is the initiator of compromise – eliminating the need for an air strike – it is not surprising that this is what happened.

To sum up, the Theory of Moves renders game theory a more dynamic theory. By postulating that players think ahead not just to the immediate consequences of making moves, but also to the consequences of counter-moves to those moves, counter-countermoves, and so on, it extends the strategic analysis of conflicts into the more distant future. TOM has also been used to elucidate the role that different kinds of power – moving, order, and threat – may have on conflict outcomes, and to show how mis-information can affect player choices. These concepts and the analysis have been illustrated by numerous cases, ranging from conflicts in the Bible to disputes and struggles today.

Example 13.5: Monopoly Venture

This game represents a firm choosing whether or not to compete in a monopolistic market. There are no competitors if they enter the market. If the firm does not enter, they get a payoff of 4, assuming 0 for the second firm so (0,4). If they do enter then a second firm, has a decision whether to fight the monopoly with the value of –2 (–2, –2) or accommodate the monopoly then prices may drop and the values are (4, 2). We see this in Figure 13.6.

If we assume prefect information and they both do a proper market study, Firm 1 compares (4, 1) to (0, 4) and chooses to enter to obtain the larger payout.

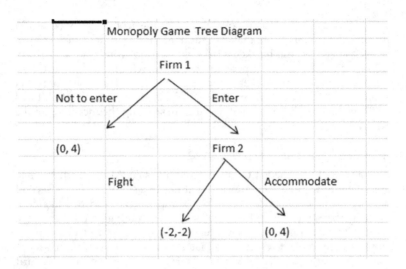

FIGURE 13.6
Monopoly game tree.

Example 13.6: Simple Politics

In an upcoming election, we have an incumbent and a challenger. Each can either run a clean campaign or a dirty campaign. In a dirty campaign, all the politician does is find and report dirt on the opponent. A possible tree diagram with values is shown in Figure 13.7.

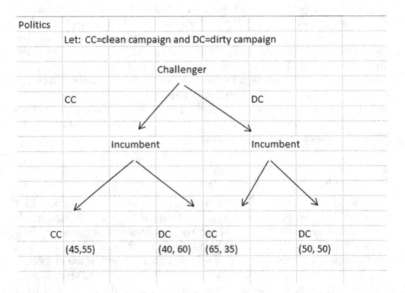

FIGURE 13.7
Political game.

Solution: If the challenger runs a clean campaign, the incumbent is better off also running a clean campaign at (45, 55). If the challenger runs a dirty campaign, the incumbent should also run a clean campaign (65, 35). The challenger is better off running a clean campaign, (45, 55) > (65, 35).

Chapter 13 Exercises

13.1 Solve the Cuban missile crisis example.

13.2 If the leader of North Koran is crazy half the time, recompute the solution.

13.3 Go back to the Run and Hide example and use the following values: $(a, b) = (5, 4)$; $(c, d) = (-2, -1)$; $(e, f) = (6, 3)$ and $(g, h) = (0, 0)$. Determine the choices made and the solution to the game.

13.4 Let's consider a three-person sequential game. Player 1 can run or hide, Player 2 can bring a friend or come alone, and Player 3 can run or walk. Payoff values are:

(run, bring a friend, run) = (−1, −1, 15)

(run, bring a friend, walk) = (−15, −5, 12)

(run, don't bring a friend, run) = (−15, 8, −10)

(run, don't bring a friend, walk) = (10, −9, 8)

(hide, bring a friend, run) = (−5, 18, −15)

(hide, bring a friend, walk) = (−20, 25, −15)

(hide, don't bring a friend, run) =(25, −5, 10)

(hide, don't bring a friend, walk) = (0, −10, 0)

Determine the choices of the players and values for the Nash equilibrium.

13.5 Two firms share the market, colluding and maintaining high prices. Each firm can decide to stop colluding and then start a price war. This decision will increase their market share and even could force the other firm to quit the market. Firm 1 can either keep colluding with Firm 2 or start a price war. If Firm 1 decides to keep colluding,

Firm 2 will need to make a decision. Firm 1 colludes and Firm 2 colludes (5, 5). Firm 1 does price war (4, 3), Firm 2 starts price war (3, 4). What should they do?

Additional Readings

Brams, S.J. (1994). *Theory of Moves*. Cambridge, UK: Cambridge University Press.

Brams, S.J. (1997). Game theory and emotions. *Rationality and Society*, 9 (1), pp. 93–127.

Brams, S.J. (1999). Modeling free choice in games. in Edited by Myrna H. Wooders. *Topics in Game Theory and Mathematical Economics: Essays in Honor of Robert J. Aumann*. Providence, RI: American Mathematical Society, pp. 41–62.

Brams, S.J. & C.B. Jones. (1999). Catch-22 and King-of-the-Mountain Games: Cycling, frustration and power. *Rationality and Society*, 11, (2), pp. 139–167.

Wilson, S.J. (1998). Long-term behaviour in the theory of moves. *Theory and Decision*, 45 (3), pp. 201–240.

Answers to Selected Exercise Problems

Chapter 2 Exercises

2.1 86.3333

2.2 94.0476

2.3 1.74

2.4 $9,200

2.5 Since E[X] = 27.50 and it is greater than 0, we should sell the policy.

2.8 (a) E[S] = 6,500 E[B] = 5,750 E[S]>E[B]. Stocks are better.

(b) $10{,}000*p - 4{,}000*(1 - p) = 7{,}000*p + 2{,}000*(1 - p) \rightarrow p = 2/3$ and $(1 - p) = 1/3$ E[S] = E[B] = 5,333.3333

2.11 US Army market survey

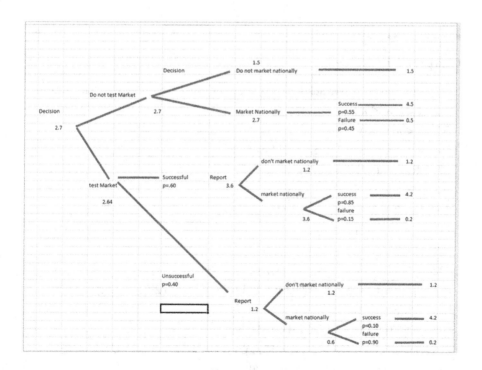

Chapter 3 Exercises

3.6 $x = 3, y = 8, Z = 100$, since reduced costs of $1 > 00.02$ so more of constraints 2.

3.7 $x = 26.66666, y = 20, Z = 120$

3.9 $x1 = 2.692308, x2 = 13.846154, Z = 303.8462$

3.10 $x1 = 3, x2 = 9, Z = 210$

3.13 $x1 = 2, x2 = 0, x3 = 8, Z = 280$

3.14 $x = 2, y = 4, Z = 110$

3.15 $x = 2, y = 4, Z = 110$

3.16 $x = 1, y = 1.5, Z = 20.5$

3.17 $x = 180, y = 260, Z = 56,400$

3.18 $x1 = 0, x2 = 6, Z = 12$

Chapter 5 Exercises

5.1 Colonel Sotto Attack City 1, Colonel Boltto, Defend City 1 (10, −10)

5.2 $a < b$ and $a > c$

5.3 $b < a, b > d$

5.6 R1C1 and R1C2 (0.5, −0.5) tie

5.7 (R2,C2) (0.275, −0.275)

5.8 R1C2 (40, −40)

5.9 R2C3 (55, −55)

5.11 R2C1 (3, −3)

5.21 R1C2 (4, −4)

5.22 4/9 R1 5/9 R2 5/9 C1 4/9 C2 (2/9, −2/9)

5.23 0.74194 R1 0.25806 R2 0.51613 C1 0.48387 C2 (0.291, −0.291)

5.24 R3C3 (4, −4)

5.25 ½ R1 0 R2 ½ R3 1/2C1 0 C2 ½ C3 (7, −7)

5.28 0.57143 R1 0.42857 R2 0.285714 C1 0.714285 C2 (1.142857, −1.142857)

5.29 4/5 R1 1/5 R2 3/5 C1 2/5 C2 (0.26, −0.26)

5.31 0.4 R1, 0.6 R2 0.9333 C1 0 C2 0.0666 C3 (1.6, −1.6)

Chapter 6 Exercises

Nash Equilibriums are (4,3) and (3,4). Neither is achievable. Trying to get to the their beat solution they could end up at (2,2).

Chapter 7 Exercises

7.1 a) Choose x b) choose x c) choose ½ u, ½ v

Chapter 8 Exercises

8.2 No pure strategy and Nash Equilibrium by equalizing strategies is (3/2, ¾) and is not Pareto optimal

8.3 (2,1) and (1,2) and not Pareto Optimal

Chapter 9 Exercises

3. (a) No pure strategy solution

Chapter 10 Exercises

10.1 ANALYSIS FOR STRATEGIC MOVES
- Simultaneous without Communication
  By movement diagram
  Conclusion: The likely outcome without communication (2, 3)
- With Communication (Strategic Moves) from Rose's Perspective

- FIRST MOVES
  Should Rose move first:

  > If Rose does R1, then Colin does C1, implying outcome (2, 3)
  >
  > If Rose does R2, then Colin does C1, implying outcome (1, 4)
  >
  > So Rose would choose outcome (2, 3), which is the equilibrium

  Should Rose **force** Colin to move first:

  > If Colin does C1, then Rose does R2, implying (2, 3)
  >
  > If Colin does C2, then Rose does R2, implying (4, 2)
  >
  > So Colin would choose (2, 3)

  Conclusions: Rose moving first would result in outcome (2, 3)
  Forcing Colin to move first would result in outcome (2, 3)
- THREATS: Example: Suppose Rose wants Colin to play C2. Can't
- PROMISES: Example: Suppose Rose wants Colin to play C2. Can't
- COMBINATION THREAT AND PROMISE: None
- Summary of Strategic Moves available to Rose (and to Colin) cannot move from equilibrium (2, 3).

Chapter 11 Exercises

11.1 No pure strategy solution. Equalizing strategies are 0.6 R1 0.4 R2 0.75 C1 0.25 C2 (2.75, 4.6) equilibrium.

Prudential Strategy leads to security levels (2.75, 4.6). Prudential Strategies are 1/8R1 7/8 R2 4/5 C1 1/5 C2. Nash Arbitration is (4.175, 5.55) when (2,7) is played 63.75% and (8,3) is played 36.25% of the time during arbitration.

Chapter 12 Exercises

12.1 Pure strategy solutions at R2C2D1 (−2, 3, −1) and at R1C1D2 (2, 4, −2).
Coalitions results
(−0.04, 0.72, −0.68)
(4, −5, 1)
(0.77, 3.89, −4.67)

Chapter 13 Exercises

13.5 (5, 5) is the Pareto optimal solution with both firms colluding.

Index

Printed in the United States
by Baker & Taylor Publisher Services